The authors—all economists—are associated with Resources for the Future's Center for Energy Policy Research. Joel Darmstadter, a fellow, has been with RFF since 1966. He is the author or coauthor of several earlier books on energy. Joy Dunkerley is a senior research associate. She was previously a staff member of the Ford Foundation Energy Policy Project. Jack Alterman, consultant in residence, was for many years assistant commissioner of the U.S. Bureau of Labor Statistics in charge of its research in input–output analysis and long-term economic growth.

How Industrial Societies Use Energy

Joel Darmstadter, Joy Dunkerley, Jack Alterman

How Industrial Societies Use Energy
A comparative analysis

Published for Resources for the Future
By The Johns Hopkins University Press
Baltimore and London

Originally published, 1977
Second printing, 1979

Library of Congress Card Number 77–83780
ISBN 0–8018–2041–3

FOREWORD

RESOURCES FOR THE FUTURE is probably not alone among research institutions in asking itself every so often why it is pursuing a particular study. What *should* it be doing? Where lies its comparative advantage? What sets the agenda?

It is not always easy to come up with good answers. Timeliness of issues, background and motivation of researchers, opportunities to raise funds, the chance to fill an obvious gap in knowledge—these and other elements go into the research menu. For this study the answer is easy. We need only recall that its in-house label never was the rather lumbering title the book now bears, but the more revealing, if somewhat frivolous, "Why can't we be like the Swedes?"

The quick answer, of course, is: because we are not the Swedes. The more thoughtful answer is that the North American continent differs from the Scandinavian peninsula in natural features such as climate and topography, and even more so in man-made ones, some hard to change in a hurry, such as housing and city patterns, dispersion of population, transportation modes, industry mix, and some more easily changed, such as the technical efficiency of energy-using devices.

Early critics of our decision to pursue the subject pointed out that the answers were obvious and could be written down on the back of an envelope: of course, countries differed in a multitude of respects, but after all, it could all be neatly summarized in the fraction that relates energy used to goods and services produced—Btu/GNP. And that ratio was demonstrably much higher in the United States than in Sweden, France, Italy, Japan, and other industralized countries.

About three years and 500 manuscript pages later we conclude that the critics, just like the endlessly repeated media characterizations of the United States as the ugly energy duckling, had greatly oversimplified the matter. There exist not only complex methodological issues indispensable for making international comparisons, such as choosing appropriate currency conversion rates, but, in addition, the task of going beyond the Btu/GDP fraction, which reveals differences but masks the reasons, proved difficult, but rewarding.

By focusing on segments of the economy and relating energy use to specific relevant activities, by distinguishing differences in consumption due to structure from those due to intensity, by tackling the problem not only on a sectoral basis but with sophisticated input–output techniques—all the while striving to keep the argument accessible to generally concerned readers—the study has opened up to serious investigation an area hitherto the preserve of instant experts. In addition, the findings should provide a sense of perspective and priority to policy makers. It leaves room for much subsequent research, especially in the area of what constitutes and brings about "structural" differences, and equally, how government policies, often not directed at energy issues at all, have contributed to bringing about the differences that the book discusses. A follow-up study has already been initiated at RFF, focusing on the development of energy/output relationships in selected countries over the past two decades. But as the first extensive study of the problem we think this book offers the reader sufficient food for thought, and the Electric Power Research Institute that funded it sufficient return on its investment.

October 1977   Hans H. Landsberg

ACKNOWLEDGMENTS

WE SHOULD LIKE, first of all, to acknowledge the collaboration of the Office of Economic Growth, U.S. Bureau of Labor Statistics, for assistance in developing the input–output estimates of energy requirements in chapter 7. BLS members who participated in this analysis were Robert A. Sylvester (project director and consumption purchases), Karen J. Horowitz (investment), David S. Frank (input–output computer processing), and Vivian Minor (statistical calculations). In addition, Arthur J. Andreassen and Valeria A. Personick did a considerable amount of work on other aspects of the input–output analysis. Ronald Kutscher, Assistant Commissioner, Office of Economic Growth, provided general supervision of the BLS work on this project.

A large number of people reviewed an earlier version of the entire study. We must record our special gratitude to Lincoln Gordon of RFF and Lee Schipper of the University of California, Berkeley, for particularly extensive comments on the manuscript. We also wish to thank the following for their helpful suggestions and criticisms: Anne P. Carter, Brandeis University; Andres Doernberg, Richard Goettle and colleagues at the Brookhaven National Laboratory; William D. Hermann, Standard Oil Company of California; Gerald Leach, International Institute for Environment and Development; Charles F. Luce, Consolidated Edison Company of New York; Gerald Manners, University College, London; Joseph Parent, Institute of Gas Technology; Henry M. Peskin, RFF; Janez Stanovnik and his colleagues at the U.N. Economic Commission for Europe, Geneva; Chauncey Starr, Electric Power Research Institute; Samuel Van Vactor, International Energy Agency; and Robert Williams, Center for Environmental Studies, Princeton University. Rex Riley, our staff monitor at EPRI, provided helpful guidance throughout the project.

Our task would have been much more arduous without valuable help on data problems from Margaret Thoroddsdottir, International Energy Agency; R. Ovart, Organisation for Economic Co-operation and Development; and Takao Tomitate and colleagues, Institute of Energy Economics, Tokyo.

We benefited from diligent research assistance at various stages of the project from David Teitelbaum, Phillip Ellis, and Linda Sanford. The processing of a manuscript of this size and complexity makes intensive and often unreasonable demands on secretarial staff. Helen-Marie Streich and Debra Hemphill performed this task with admirable skill and cooperation. Finally, Sally Skillings had the responsibility of editing a multiauthored manuscript into a unified and finished book—a job which she performed with exceptional capability and good humor.

While our study was financed by the Electric Power Research Institute, findings and conclusions are solely those of the authors.

Joel Darmstadter
Joy Dunkerley
Jack Alterman

CONTENTS

Tables

Figures

PART I

Introduction and Quantitative Setting

CHAPTER I

Why International Comparisons?

A FEW YEARS ago the question with which this study deals might have engaged the attention of a contributor to an obscure technical journal; today it has become an issue vigorously debated by economists, environmentalists, and policy makers. Why is per capita consumption of primary energy resources[1] so much higher in the United States than in other advanced industrial countries—such as Sweden, West Germany,[2] and France—whose per capita income does not differ appreciably from that of the United States? The importance of the issue is clear enough. If the answer to the question points to the possibility that the American standard of living can be maintained at a much lower level of energy use, such a finding could have major significance for the direction and eventual payoff from conservation strategies that the United States might see fit to undertake.

What we offer in this study is a first step toward answering this question. It is a quantitative reconnaissance in which we depict the comparative patterns of energy consumption for nine countries that are considered to have high incomes by international standards. These countries are: United States, Canada, France, West Germany, Italy, Netherlands, United Kingdom, Sweden, and Japan. Further, we identify those components of energy consumption that give rise to variations between countries in the relationship between energy use and national output. Finally, we interpret, as far as possible, the respective contributions of (1) economic structure and (2) characteristics of energy utilization to the

---

[1] Primary energy resources are comprised of fossil fuels, hydroelectric power, and nuclear power; when refined or processed into products like gasoline or electricity, they constitute the forms in which energy is used in daily life.

[2] The Federal Republic of Germany will be referred to as West Germany throughout this study.

intercountry variations in energy and output. Where the data permit, underlying factors—such as relative price differences and demographic features—are woven into the analysis. Policy aspects and lessons for the United States, along with other considerations that were found to be fairly elusive, enter into the discussion less often. This is not because such questions are less important. They are, of course, vitally important, but in view of the fact that study in this field has barely begun, we felt that an effort to lay out facts and disentangle quantitative relationships was a worthwhile and sufficiently complex task to be the central focus in this initial undertaking.

Relationships Between Energy and Output

The starting point for our subject involves a dual set of quantitative phenomena. First, there is an unmistakably strong cross-sectional, multi-country correlation between national output[3] per capita, on the one hand, and energy consumption per capita, on the other. (See figure 1-1.) A high positive correlation between national output per capita and energy per capita does not, however, signify a one-to-one linkage between the two: a given percentage increase in energy need not be associated with the same percentage increase in output. Instead, that relationship is established by the slope of the regression line, which for countries of over $1,000 in per capita national output, typically shows a regression coefficient of about 0.8; that is, a 10 percent rise in output per capita is associated with an 8 percent rise in energy used per capita. (But note that the coefficient in figure 1-1, which includes numerous countries with a lower per capita GDP, is 1.0.)

There is a historical counterpart to the "snapshot" plot in figure 1-1. It shows that, for given countries, changes in per capita output over time—measured, throughout this study, in dollars of constant 1972 purchasing power—closely parallel changes in per capita energy utilization. Figure 1-2 illustrates this point with respect to four of the countries in our study over a recent time span. A longer time series would dramatize the association even more. The reason for the cross-sectional and historical patterns is neither unexpected nor particularly subtle. Energy utilized in productive activity is one of the components of economic growth, just as proceeds of that economic growth and rising income permit the consumption of energy-associated creature comforts and other services.

---

[3] Expressed as gross domestic product (GDP) throughout.

FIGURE 1-1   National Output Per Capita Versus Energy Consumption Per Capita for Selected Countries, 1972

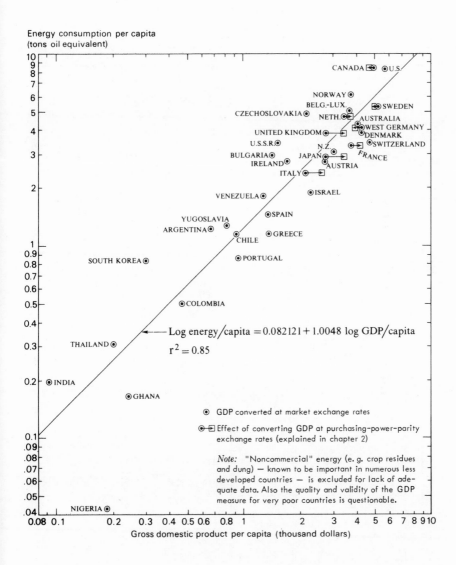

Energy consumption per capita
(tons oil equivalent)

CANADA ⊞● ● U.S.

NORWAY ●
BELG.-LUX. ● ⊞● SWEDEN
CZECHOSLOVAKIA ● NETH.●⊞ AUSTRALIA
⊞● WEST GERMANY
UNITED KINGDOM ●—⊞ ●⊞ DENMARK
U.S.S.R. ● N.Z. ●—⊞ ● SWITZERLAND
BULGARIA ● JAPAN ●—⊞ *FRANCE*
IRELAND ● AUSTRIA
ITALY ●—⊞

VENEZUELA ● ● ISRAEL

● SPAIN
YUGOSLAVIA ●
ARGENTINA ● ● GREECE
CHILE

SOUTH KOREA ● ● PORTUGAL

● COLOMBIA

Log energy/capita = 0.082121 + 1.0048 log GDP/capita
THAILAND ●
$r^2 = 0.85$

● INDIA

● GHANA

⊙   GDP converted at market exchange rates

●—⊞ Effect of converting GDP at purchasing-power-parity
exchange rates (explained in chapter 2)

Note: "Noncommercial" energy (e. g. crop residues
and dung) — known to be important in numerous less
developed countries — is excluded for lack of adequate data. Also the quality and validity of the GDP
measure for very poor countries is questionable.

NIGERIA ●

Gross domestic product per capita (thousand dollars)

Whether we are talking about mechanized industrial or agricultural activities, freight or passenger transport, or the illumination and heating of
commercial and residential structures, the application of fuels and power
is clearly a critical element in the state of a nation's economy—one
which, not surprisingly, is distinctly related both to levels of development
and economic change over time.

FIGURE 1-2   *Energy Consumption Per Capita Versus National Output Per Capita for Four Selected Countries, 1961–1974*

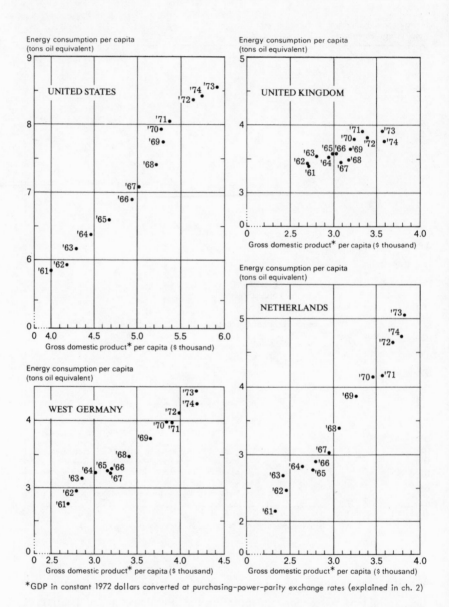

*GDP in constant 1972 dollars converted at purchasing-power-parity exchange rates (explained in ch. 2)

But there is a second interesting phenomenon emerging from the scatter of points in figure 1-1, and it is this feature which gives rise to the present study: countries with similar output per capita vary widely in per capita energy consumption. It should be noted that for purposes of this study, when discussing output we employ the national accounting measure of gross *domestic* product (GDP) rather than gross *national* product (GNP). For most countries, the difference between the two indicators is quantitatively minor, but GDP has the virtue of covering only a nation's domestic economic activity, excluding net factor income originating in overseas enterprises and investments, and is therefore the more appropriate national accounts measure to which to relate a nation's domestic energy consumption.

The wide variation in energy consumption can be seen by the fact that Sweden's 1972 per capita GDP was about 11 percent below the U.S. level, but Sweden's per capita energy consumption was 36 percent below that of the United States. France and Germany were also disproportionately low per capita consumers of energy when comparing their per capita GDPs with that of the United States. But Canada's level of output per capita was lower than the U.S. level, while its level of per capita energy consumption was practically equivalent to that of the United States.

TABLE 1-1   Energy/Output Relationships, 1972

Country	GDP per capita (dollars) [a]	Energy consumption per capita (tons oil equiv.) [b]	Energy/GDP ratio (tons oil equiv. per $ million)	(Indexes, U.S. = 100)
United States	5,643	8.35	1,480	100
Canada	4,728	8.38	1,772	120
France	4,168	3.31	795	54
W. Germany	3,991	4.12	1,031	70
Italy	2,612	2.39	915	62
Netherlands	3,678	4.68	1,272	86
United Kingdom	3,401	3.81	1,121	76
Sweden	5,000	5.31	1,062	72
Japan	3,423	2.90	849	57

Sources: Energy and GDP data are from chapter 3; a description of methods used in converting GDP data to U.S. dollars through the use of purchasing-power-parity ratios appears in chapter 2 and in appendix A "Derivation of GDP Estimates"; raw energy data are from Organisation for Economic Co-operation and Development, *Statistics of Energy;* and basic GDP data from OECD, *National Accounts of OECD Countries.*

[a] Foreign currencies were converted into dollars using exchange rates which reflect comparable purchasing power. Chapter 2 and appendix A provide a full explanation.

[b] Tons of oil equivalent is the unit of energy measurement generally adopted for this study. (One million tons equal roughly 20,000 barrels per day.)

FIGURE 1-3   Energy/Output Ratios Versus National Output Per Capita for Selected Countries, 1972

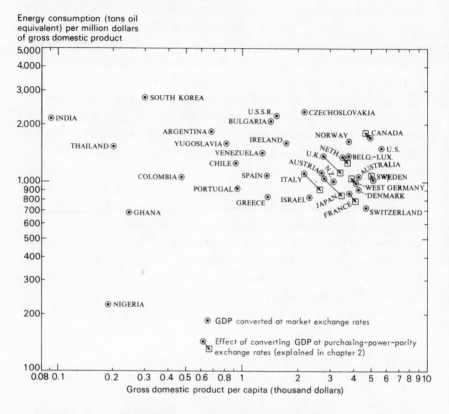

Energy consumption (tons oil equivalent) per million dollars of gross domestic product

A salient way of exposing such intercountry differences is to focus on the ratio of energy consumption per capita to GDP per capita (expressed in shortened form as energy/GDP, or energy/output). It is this energy/ GDP ratio (population canceling out in the calculation) that, statistically speaking, occupies center stage in our study. In table 1-1 on the preceding page is a summary of energy/output relationships.

The ratios of energy to GDP for these and other selected countries also appear in figure 1-3. Here, the ratios are plotted against per capita GDP in order to reveal graphically some additional characteristics of the relationship between energy and output. The figure shows considerable disorder in this representation—in sharp contrast to the relationships portrayed in figure 1-1. There is no hint of a clear-cut correlation— positive or negative—between energy/GDP ratios and per capita output. For countries whose per capita GDP exceeds the $900–$1,000

FIGURE 1-4   Energy/Output Ratios for Five Selected Countries, 1961–1974

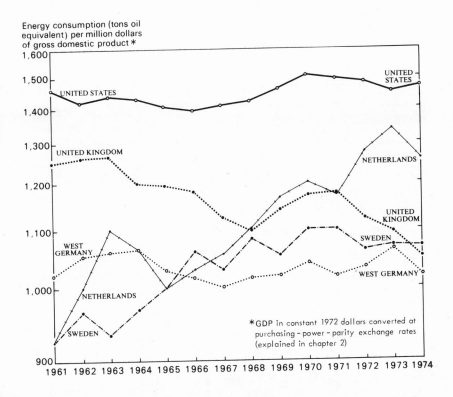

Energy consumption (tons oil equivalent) per million dollars of gross domestic product *

*GDP in constant 1972 dollars converted at purchasing-power-parity exchange rates (explained in chapter 2)

range, there is a slight tendency for the ratio to decline as income rises, although it happens that in the group of nine high-income countries in our study, the opposite situation seems to obtain. There is, however, a wider range of values for the very poorest countries than for the very richest countries. (Problems of data reliability may contribute to the dispersion of points in the former case.)

A reminder, finally, that cross-sectional patterns exist within the dynamics of historical change is shown in figure 1-4, which illustrates developments in the energy/GDP ratios for five of the high-income countries over roughly a decade and a half. No generalizable trend is apparent: sharply rising ratios for Sweden and the Netherlands; a declining trend for Britain; and a fluctuating, but no decisive up or down movement for either the United States or West Germany. (However, over a longer historical time span—in the U.S. case between the 1920s and 1950s—the ratio drifted downward for both countries.) Such variability in trends underlines a shortcoming of the cross-sectional, snapshot approach that studies consumption patterns in one year only (figures 1-1 and 1-3).

The Public Debate

Perhaps it is precisely the absence of systematic explanations for levels or movements in these ratios that has prompted numerous persons to question the "inevitability" of a ratio uniquely applicable to a given country at a point in history; and, more specifically, to ask whether the lower per capita energy consumption accompanying demonstrably high living standards in many advanced countries might not be a fitting model for the United States to emulate as it goes about its search for energy-conserving possibilities and policies.

Comments on this topic have not been noticeably beset by self-doubt. Thus, Paul Ehrlich:

> Beautiful, peaceful, democratic Denmark uses 48 percent as much energy per capita as we do. . . . The case of Sweden is especially instructive. By the measure much beloved of the growth maniacs, per capita GNP, this heavily industrialized nation in a cold climate has a standard of living slightly better than the United States. Yet, it achieves this superiority while consuming about half as much energy per capita as we do. . . . Countries like Sweden are simply cleverer than the United States in extracting more benefit from less energy.[4]

Given the premise that the level of U.S. energy consumption is not warranted by the level of U.S. national output, it is a short and logical step to conclude that a high ratio of energy consumption in the United States is, by the yardstick of other countries, symbolic of waste: "energy squandered by 205 million Americans equals the energy used by 105 million Japanese."[5] Or, from John Sawhill, the former Federal Energy Administration administrator, commenting on America's emergence as an "energy extravagant" society, ". . . with only 6 percent of the world's population, we use 35 percent of the world's energy, and not merely because U.S. society is so highly developed. Fully 30 percent of our

[4] Paul R. Ehrlich, "An Ecologist's Perspective on Nuclear Power," *Public Interest Report* vol. 28, no. 5-6 (May-June 1975) p. 3.

[5] Thomas O'Toole, "U.S. Waste of Energy Remains Vast," *The Washington Post,* February 24, 1975, p. A1.

energy use is pure waste."[6] Even if our future energy growth were to be little more than half of our national output growth, that achievement would, in Dr. Sawhill's judgment, be no ground for self-congratulation in the light of accomplishments in Sweden, Denmark, and West Germany. For at "that rate it would take 30 years before we approached the efficiency already achieved by the three countries. . . ."[7]

Others, far from seeing the scope for considerable flexibility in the amounts of energy compatible with given quantities of material output, tend to seize upon energy consumption and output as activities that are almost rigidly circumscribed in their relationship to each other. One such argument is contained in a recent publication of the Chase Manhattan Bank. Ignoring a progressive decline in the U.S. energy/output ratio occurring during three decades after the mid-1920s, the bank nonetheless believes that the roughly parallel trend of more recent years augurs an enduring constancy in the relationship between energy and output. "There is no sound, proven basis for believing a billion dollars of GNP can be generated with less energy in the future." Similarly, the international record, as viewed through the Chase Manhattan lens, emerges in equally absolutist—and, indeed, in factually erroneous—terms: "Among the different areas of the world, there is a wide variation in the per capita use of energy. And, significantly, there is an almost identical variation in the per capita GNP."[8]

Energy Intensity and Composition of Output

That such statements reflect a tone of hyperbole is no ground for dismissing them. The fact is that there simply has been no solid empirical basis for informed judgments about reasons for international variability in the relationship between energy and output. Is such variability primarily the consequence of the energy intensity that characterizes given activities within economies? That is, is one country's wheat or steel production more energy intensive than another's? Or does the variability arise from the fact that the relative importance of given energy-using activities varies

6 John C. Sawhill, remarks at "Project Independence" hearing, San Francisco, October 7, 1974, Federal Energy Administration press release.

7 John C. Sawhill, statement before the Joint Economic Committee, 93 Cong. 2 sess. (November 19, 1974).

8 John G. Winger and Carolyn A. Nielsen, "Energy, the Economy, and Jobs," *Energy Report from Chase*, September 1976, pp. 2–3.

from country to country—in other words, that there are important structural (or output mix) differences among the countries? That is, does one country specialize in activities such as steelmaking that generally use a lot of energy, while another pursues activities such as agriculture, which tend to be less energy intensive? The respective roles of these two factors—the first, energy intensity, the second, structure—are an indispensable, and probably minimal, part of conducting an intercountry analysis of energy consumption patterns, because aggregative comparisons of the relationship between energy and output can obscure aspects highly relevant to the analysis. Thus it is possible for a country to have a much higher energy/GDP ratio than another simply because of the predominance of inherently energy-intensive activities.

Take the following simple table in which E = units of energy; O = units of output, or GDP; and E/O = the energy/GDP ratio.

Sectors	Country A			Country B		
	E	O	E/O	E	O	E/O
Agriculture	25	50	0.50	120	120	1.00
Steel	350	200	1.75	60	30	2.00
Total	375	250	1.50	180	150	1.20

The table gives hypothetical energy and output figures for two countries, each assumed to be entirely composed of two sectors—agriculture and steel. The central question to be answered is why country A's E/O ratio (1.50) exceeds that of country B (1.20). Or, as it is sometimes posed, why is country A, in the aggregate, more "energy intensive" than country B? In fact, note that in both agriculture and steel, country A is less energy intensive than country B. But steel is seen to occupy a disproportionately large place in country A's economy—80 percent of total output, while steel's place in country B is only 20 percent. Consequently, since energy intensity in steelmaking is higher than in agriculture, it is clear that the emphasis on steel production in country A produces a higher aggregate E/O ratio than in B, despite A's lower energy intensity in both sectors. The structure–intensity dichotomy is one on which we focus recurrently in this study.

Related Questions

Of course, both structural and energy-intensity characteristics of an economy may be only the surface manifestations of more fundamental features—geography, resource endowment, technology, demographic factors, and economic policies—all of which an ideal multicountry analysis should include. For instance, a comparative advantage in steelmaking (including the ability to export steel to countries lacking that capacity) may have been conferred upon a country through abundant coal and iron ore deposits—circumstances that have little to do with so-called energy profligacy. A related point concerns contrasting mixes of basic fuels prevailing in different countries. The efficiency with which various fuels can be used differs—at least to an extent—even when used for the same purpose. Therefore, different fuel mixes can lead to different national energy/output ratios. On the other hand, the extent to which a country's overall energy intensity is influenced by, say, the prevalence or absence of large cars with poor operating characteristics may reflect a government's inclination or disinclination to employ aggressively such economic instruments as fuel and horsepower taxes.

It is also worth noting that energy-intensity differences among countries in a given activity may, but need not, spell efficiency differences in the economic use of *all* the resources going into that activity. It is only partially revealing, in other words, to point to a country's lower (or higher) energy intensity in a given pursuit without specifying the balance of economic advantage in terms of total resource cost. However, such an overall reckoning requires knowledge about the relative requirements not only for energy but for capital, labor, and other inputs whose costs—in most economic activities—may swamp energy costs by a factor of 25 to 1. Yet, because that aspect constitutes a complex research undertaking of its own, this broader dimension is largely missing from our study—which is not one that analyzes total economic efficiency.

Finally, there is the need to say something about the appropriateness of choosing GDP (and its components) as the quantitative reference point for comparative energy analysis. GDP and GNP are, of course, conventional yardsticks of a country's market-oriented economic activity. The use of GDP in this study does not reflect confidence that growth in such a measure necessarily produces enhanced human welfare or perceived happiness, or that a shrinking national product need signify an erosion of such welfare. Gross domestic product and its constituent parts *do,* however, represent an objective reckoning of expressed market preferences, whereas most other standards for measuring the social product would almost certainly involve the imposition of intensely controversial and debatable individual value judgments.[9] Such questions are exceedingly important elements in the contemporary debate over economic growth, environment, resource scarcity, and life-styles. We hope the results of this study—built on presently available, concrete, and quantifiable topics—can take their place with approaches emphasizing other standards of relevance.

As the purpose of this study is to analyze variations in energy consumption among highly industrialized countries, only those countries with relatively high levels of output per capita (about twenty in number) were considered for inclusion. From this group, nine were selected. We considered nine a sufficient number to permit generalized analysis, yet still small enough to allow the degree of detail we felt essential to the study.

Within the group of twenty relatively high-income countries, the nine selected were those countries with the largest economies (United States, Japan, West Germany, France, United Kingdom, and Italy); a small country (the Netherlands); a geographically large country (Canada); and a country with a large proportion of hydroelectricity (Sweden). These nine span a wide range of energy/GDP ratios. The highest, Canada, is over twice that of the lowest, France. The main group of industrialized countries not represented are the centrally planned economies.

However, because of limited coverage in the underlying source document, it was not possible to conduct the analyses of chapters 7 and 8 for all nine countries. That part of the study includes United States, France, West Germany, Italy, United Kingdom, and Japan.

The year 1972, applicable to most of the study, was chosen not only because it was the most recent year for which sufficiently detailed data were available, but mainly because it preceded the sharp increase in the price of crude oil, the disruption caused by the oil embargo, and the subsequent and protracted recession. A study based on 1972 could therefore be expected to give a picture of energy consumption under conditions of high economic activity. The analysis in chapters 7 and 8, however, applies to the year 1970.

---

[9] Work now being done at the National Bureau of Economic Research and elsewhere deals with objective modification of conventional output and income measures to make them reflect welfare more closely.

In chapter 2 we introduce the reader to some of the principal conceptual and measurement problems encountered in a comparative study of this kind. Chapter 3 surveys comparative energy consumption patterns—in the aggregate, and in sectoral composition. Chapters 4, 5, and 6 constitute sectoral approaches in which structural and intensity characteristics (along with other fundamental aspects) are applied, respectively, to the household-commercial, transport, and industrial segments of the economy. Chapters 7 and 8 rely upon the recently completed ICP study, which provides multicountry comparisons of GDP and components and data on price and quantity differences.[10] Specifically, chapter 7 uses an input–output technique to provide insight into the roles of structure and energy intensity in shaping overall energy/GDP variability among countries. Chapter 8 deals with the energy and energy-related components of consumer purchases in the light of differences in relative prices. In chapter 9, the significant findings of the study are brought together in an interpretive setting.

[10] Irving B. Kravis, Zoltan Kenessey, Alan Heston, and Robert Summers, *A System of International Comparisons of Gross Product and Purchasing Power* (Baltimore, published for the World Bank by The Johns Hopkins University Press, 1975).

CHAPTER 2

Problems and Definitions in the Sectoral Approach

THIS STUDY uses a two-pronged approach to probe the underlying factors accounting for intercountry variation in energy consumption and output. The first approach is an examination of specific functional sectors of the countries involved. The second is application of a final-demand, input–output mode of analysis, involving broad comparisons of relative prices, income, and total and disaggregated expenditures. The purpose of this chapter is to set the scene for the sectoral analysis. As a first step we define with some precision "energy consumption" and "gross domestic product"—terms which are used to compare the energy intensities of different economies. In dealing with a subject as controversial as this, it is essential to be quite clear about what is included (and under what conditions), and what is not included. Right from the start, some of the apparent differences in energy intensities arose from variations in definitions. In the case of both energy consumption and gross domestic product, problems of comparability were encountered. We shall describe here only the major problems, giving some idea of the order of magnitude and the direction of difference involved; that is, attention will be drawn only to those definitional problems that could lead to a significant widening or narrowing in energy intensities between countries, particularly between United States and the other countries.

Problems in Measuring Energy Consumption

Standardization of Definition. Some degree of standardization in both definition and coverage of energy consumption was assured by basing our statistical analysis on energy data published by the Organisation for

Economic Co-operation and Development (OECD).[1] These data are organized within a framework of consistent and standardized definitions that greatly facilitate intercountry comparison. For example, for all countries the basic definition of energy consumption is the same.[2] It covers total energy supplies (whether from national production or from imports) minus all quantities not consumed in the country (whether exports or ship and aircraft bunkers). The whole is adjusted for relatively small stock changes. In fuel coverage, too, there is equal uniformity. Data for all countries cover the same primary fuels—coal, natural gas, crude petroleum, and primary electricity—and a similar range of secondary fuel products. Noncommercial fuels such as wood and charcoal are not included. Even within these consistent definitions, however, there is room for some ambiguity that can distort comparisons between countries.

Conversion Factors. For example, data expressed in tons of coal, cubic meters of gas, or kilowatt hours (kWh) of electricity must be converted into a common unit of thermal measurement to permit comparison between countries. The unit adopted for this study is million tons oil equivalent (mtoe), roughly equal to 20,000 barrels per day.[3] A well-established series of conversion conventions exists for this purpose, but even so, some difficulties remain. The thermal content of coal, for example, has usually been assumed to be the same for all countries even though it is well known that the actual thermal content of coal varies considerably among countries. This matter is of particular relevance to the United States and the United Kingdom, both of which use large quantities of coal of somewhat lower thermal content than in other countries. Adjustment to take this into account reduced the U.K. energy consumption by 5 percent and U.S. by just under 2 percent.

[1] The statistical data in this chapter and in chapters 4, 5, and 6 are based on the country balance sheets for 1972 in Organisation for Economic Co-operation and Development, *Statistics of Energy 1959–72* (Paris, OECD, 1974) (see appendix B of this study). An account of the conversion procedure and the method employed in making country balance sheets expressed in million tons oil equivalent (mtoe) from which the appendix B tables were drawn is available from the Energy and Materials Division of Resources for the Future. A recent OECD publication, *Energy Balances of OECD Countries 1960–74* (Paris, 1976), performs essentially the same operation and contains data similar but not always identical to those appearing in this study.

[2] "Total inland consumption" as defined by OECD. For convenience, referred to in this study as "total energy consumption."

[3] The tons oil equivalent (toe) basis of energy measurement has been adopted for this study because it is the unit customarily used by OECD—the source of many of our basic data. One million ($=10^6$) oil-equivalent tons represent approximately 1½ million coal-equivalent tons or 40 trillion ($=10^{12}$) British thermal units (Btus) or 10 trillion kilocalories.

Treatment of Hydroelectricity. Another source of ambiguity in conversion conventions concerns the converting of hydroelectricity to a thermal basis. Several possibilities were available, ranging from the conversion of hydroelectric generation as if it incurred no losses (that is, 100 percent efficiency), to the conversion of hydroelectricity as if it incurred the same losses as fossil fuel generation (30–35 percent).[4] It will be appreciated from this wide range, that the choice of efficiency can make a considerable difference to energy consumption for those countries— Canada and Sweden—in which hydroelectricity represents a large proportion of total energy generation. Canada's energy consumption, for example, is 16 percent greater with the use of a lower rather than a higher efficiency, though in both cases, Canada remains a high consumer. Sweden is somewhat different. Energy consumption could be increased by over 20 percent depending on the hydroelectric conversion used, and such an adjustment makes a considerable change in the position of Sweden relative to other countries. From being a comparatively low consumer by European standards, it becomes among the highest—still of course remaining below the United States, but far less dramatically.

As several alternatives exist and are important for some countries, the question was which to choose. None is entirely satisfactory; all are used more appropriately in some circumstances than in others. Since our purpose was primarily the comparison of energy consumption among countries whose electric generating capacity is largely fossil fueled, we have chosen the 35-percent efficiency conversion and adjusted all input into hydroelectric generation as if that generation were based on fossil fuel generation.

Noncommercial Fuels. Another possible source of distortion arises from the exclusion of "noncommercial"[5] fuels (wood, wood wastes,

---

[4] These efficiencies do of course differ from country to country, but as we were unable to obtain satisfactory data for all countries, we standardized on the basis of U.S. efficiencies. In particular, this was done by taking the estimated heat rate of the United States for 1972 (10,479 Btu per kWh—source: Edison Electric Institute, *Statistical Yearbook of the Electric Utility Industry, 1972* [New York, EEI, 1973])— adjusted to a gross output basis (that is, including generating plants' own use of electricity), to yield an efficiency of about 35 percent (that is, a conversion rate of 10^9 kWh = 0.244 mtoe). This conversion was applied to all primary electricity. This 35 percent, which will appear high to U.S. readers, is actually low because OECD statistics—unlike those of the United States—include the 5 percent own use by power plants as part of total electric production. Note that the 35 percent does not include distribution losses, which in our format are included under consumption by the energy industries.

[5] "Commercial" and "noncommercial" may not be the most appropriate terminological distinction, because such things as wood wastes may be commercially very important in certain enterprises. The distinguishing characteristic is whether such fuels have an organized market or not.

charcoal) from our energy data. As we are dealing only with the energy consumption of highly industrialized countries, it is reasonable to believe that such an exclusion, which would be impossibly misleading if we were including poorer agricultural countries, is not of major importance. Nonetheless, for some countries, in some sectors, noncommercial sources of fuel may be significant. For example, in Canada as late as 1958, over 10 percent of total supplies of fuel used for domestic heating was in the form of wood, and even in the 1970s wood is used to generate a part—admittedly very small—of total electricity supplies. Other countries of our group (Japan, Italy, France, and Sweden) that are less heavily urbanized in some areas may consume large quantities of wood and charcoal that do not enter into our data. Certainly it is difficult to believe that Japan could subsist on the low fuel consumption in the household sector (described in chapter 4) without using a significant amount of wood products. In Sweden, industrial use of wood waste and waste liquor (by-products of paper making) account for almost 3 million tons oil equivalent per year, an amount which would increase our Swedish consumption figures by almost 7 percent.[6]

Waste Heat and District Heating. Another possibility of underestimation in the fuel and power consumption of some countries arises from the exclusion from OECD data of (1) waste heat sold commercially by power stations operated by the public service and (2) of district heating, that is, the heating of public buildings, stores, offices, and apartments by steam or hot water purchased from a district heating plant set up for that purpose. Because such deliveries result from electricity generation, the fuel used in their production will, in principle, be included in our data on total energy consumption, so that comparisons among countries on the aggregate level will not be distorted. On the sectoral level, however, some difficulties may be encountered. In Sweden, for example, some 20 percent of all heating needs are met by district heating (mainly hot water plants), while for most other countries, district heating accounts for only about 1 percent of total household consumption. At the sectoral level insofar as district heating is excluded, Swedish consumption would be underestimated relative to that of other countries.[7]

[6] The recently published OECD balance sheets (OECD, *Statistics of Energy*) now include wood waste in Swedish energy consumption.

[7] A recent detailed study of U.S. and Swedish energy consumption by Schipper and Lichtenberg that was based on direct comparison of U.S. and Swedish rather than the OECD sources, shows Swedish energy consumption to be somewhat higher than our data. This is because Schipper and Lichtenberg included wood wastes, naphtha imports used as petrochemical feedstock, and higher refinery losses. Lee Schipper and A. J. Lichtenberg, *Efficient Energy Use and Well-being: The Swedish*

With regard to energy consumption data, therefore, despite the strong element of uniformity and standardization imposed by the use of OECD data, several elements of noncomparability and ambiguity remain. The most important concern is the handling of hydroelectricity and the exclusion of noncommercial fuels. Neither of these elements is of major importance in explaining the generally lower level of energy consumption in Western Europe and Japan compared with the United States, though they do affect the position of individual countries, Sweden in particular.[8]

The Problem of Measuring Total Output

A major problem arose in the measurement of total national output (gross domestic product)—the denominator in the calculation of the energy/GDP ratio. It was how to convert output data measured in national currencies to a common unit of currency to permit intercountry comparison. The use of market exchange rates has obvious disadvantages. They are derived from the internationally traded component of national output and are therefore not well adapted to the conversion of the quantitatively much larger nontrade component. As late as 1970 it was found, for example, that $1,000, when converted to sterling at the official exchange rate, bought a basket of U.K. goods 52 percent larger than the same expenditure in the United States. To overcome this problem, an alternative set of rates reflecting the real purchasing power of each country's currency was estimated and used to convert each country's 1972 GDP to U.S. dollars.[9] (See appendix A for details of how GDP

Example (Berkeley, Calif., Lawrence Berkeley Laboratory, University of California, 1976).

[8] Note that not all energy stays within the country where it is initially consumed. Some leaves the country in the form of energy embodied in nonenergy exports; and some enters a country through energy embodied in nonenergy imports. Part of the difference in levels of energy consumption among our countries could be due, therefore, to differences in the amounts of energy embodied in nonenergy trade. This topic is discussed in chapter 6.

[9] The basis for these rates—or so called purchasing-power-parity rates—is contained in Irving B. Kravis, Zoltan Kenessey, Alan Heston, and Robert Summers, *A System of International Comparisons of Gross Product and Purchasing Power* (Baltimore, published for the World Bank by The Johns Hopkins University Press, 1975). This study, which is Phase One of the United Nations Comparison Project, was produced by the Statistical Office of the United Nations, the World Bank, and the International Comparison Unit of the University of Pennsylvania. The report will be referred to hereafter as the International Comparison Project (*ICP Report*). This study, based on exhaustive surveys of price and quantity data for a group of countries, yielded estimates of purchasing power parities for five of our countries (France, West Germany, Italy, United Kingdom, Japan) in terms of U.S. dollars for the year 1970. These were updated to 1972, and additional estimates were made for the missing countries—Canada, the Netherlands, and Sweden. Details of the estimating procedure are given in appendix A.

estimates were derived.) The use of such purchasing power parities, despite the fact that they themselves are approximate, is preferable for our purpose, which is to link energy consumption with a measure of *real* rather than nominal output of an economy.

The effect of using purchasing power parities rather than exchange rates is to raise the real output of other countries relative to that of the United States, the relative increase being more marked, the poorer the country in relation to the United States. Among our eight other countries, Canada and Sweden have recorded levels of per capita income which are already comparable to the United States, so that the use of purchasing power parities (admittedly closely based on exchange rates) does little to change their level of output relative to the United States. But our study also includes Italy, United Kingdom, and Japan whose nominal levels of per capita income are substantially below that of the United States and whose level of real output is sharply increased by the use of purchasing power parities. The real output of these countries appears some 20 percent higher than the level given by converting at market exchange rates.

Comparative Energy/GDP Ratios

Energy/GDP ratios are, therefore, affected by various methods of calculating both energy consumption and GDP. By way of illustration these are quantified in table 3-1. Rather than give all possible permutations of differently defined energy consumption and GDP, we show only the most critical: energy consumption calculated alternatively as 35- and 80-percent efficiencies for hydroelectricity, and GDP converted by both purchasing power parities and exchange rates. For some countries —France, Germany, and the Netherlands—energy/GDP ratios change very little regardless of the combination of energy consumption and GDP measures used. Canada and Sweden, however, are very sensitive to the convention used for aggregating hydroelectricity; and Italy, Japan, and the United Kingdom, to the use of real output rather than output converted at exchange rates.

For reasons developed earlier, we have based our analysis on energy/GDP ratios (see column 9 of table 3-1) calculated from energy consumption data using the lower efficiency for hydroelectricity (column 2), and GDP data (column 4) based on real output deflated by purchasing-power-parities rather than exchange rates. The effect of this choice compared with other options is to increase the energy/GDP ratios of countries which have a large amount of hydroelectricity in relation to the United

States. For example, with a given GDP, Canada increases from (U.S.= 100) 106 to 120, and Sweden from 59 to 72.

With regard to the choice of output measure, the effect on the energy/ GDP ratio of choosing the purchasing-power-parity conversion rather than the exchange rate is generally to *widen* the gap between the United States and other countries because, as stated before, the purchasing-power-parity rate tends to increase the level of real output relative to the United States. Consequently, with an unchanged amount of energy consumption, the energy/GDP ratio will therefore be lower, because the same amount of energy is associated with a larger output. As can be seen in table 3-1, the energy/GDP ratios of those countries with the lowest income per head—Italy, the United Kingdom, and Japan—all fall sharply.

CHAPTER 3

Comparative Patterns of Energy Consumption

Comparative Energy Consumption in the Aggregate

Thus, having decided to use energy data based on the aggregation of primary electricity at 35 percent and GDP converted at purchasing-power-parity rates, we now examine some of the overall patterns of energy consumption. Looking at the basic series of energy/GDP data used in this study (table 3-1), in column 9 we see that Canada is the most energy intensive of all countries, followed by the United States. The other countries fall fairly regularly, without a major discontinuity in energy intensity, to the lowest, France, whose energy consumption per unit of output is about one-half that of the United States and Canada. The energy intensities of the European countries and Japan are, with some variation between countries, on average about two-thirds of the U.S. level. That is, most European countries and Japan produce a comparable value of output using one-third less energy than do the United States and Canada.

An additional comparison is worth making here. It will be remembered that the choice of method for aggregating hydroelectricity made a large difference when comparing total energy consumption for some countries, particularly Canada and Sweden. But this problem exists primarily on the aggregate level, and is contained largely in one sector—the transformation loss sector.[1] This sector, which will be described in detail in chapter 6, accounts for the losses sustained in transforming energy from one form to another. It consists almost entirely of heat losses that are incurred in the production of electricity from fossil fuels, or, in the case of hydroelectricity, their imputed equivalent. If this sector is subtracted from the total, we are left with a new total representing energy

---

[1] This would not be the case if electricity were generated on-site rather than purchased from utilities.

TABLE 3-1 *Energy/GDP Ratios Under Varying Measures of Energy Consumption and GDP, 1972*

Countries	Energy consumption w. hydro at 80% efficiency	Energy consumption w. hydro at 35% efficiency	GDP using market rates	GDP using binary ideal ppp[a]	Energy consumption w. hydro at 80% & GDP using market rates	Energy consumption w. hydro at 80% & GDP using ppp[a] rates	Energy consumption w. hydro at 35% & GDP using market rates	Energy consumption w. hydro at 35% & GDP using ppp[a] rates	Energy consumption w. hydro at 35% & GDP using ppp[a] rates
	million tons oil equivalent		billion U.S. dollars		index, U.S. = 100				tons oil equiv./ $ mil GDP
	(1)	(2)	(3)	(4)	(5)	(6)	(7)	(8)	(9)
U.S.	1,706	1,745	1,178	1,178	100	100	100	100	1,480
Canada	158	183	106	103	103	106	117	120	1,772
France	164	171	196	215	58	53	59	54	795
W. Germany	251	254	258	246	67	70	66	70	1,031
Italy	124	130	118	142	73	60	74	62	915
Netherlands	62	62	46	49	93	87	91	86	1,272
U.K.	212	213	154	190	95	77	93	76	1,121
Sweden	35	43	41	41	59	59	71	72	1,062
Japan	296	308	294	363	69	56	71	57	849

Source: Data in appendix B.
[a] purchasing power parity.

delivered to final consuming sectors, which excludes heat losses—real or imputed. This total final consumption has therefore the advantage of being more comparable among countries in terms of the ultimate uses of energy, although it is, of course, inadequate as a measure of primary energy requirements. We nevertheless found again that most European countries and Japan produce a comparable value of output using, in the final consumption sectors, on average two-thirds of the energy that North America does. This new total did, however, alter the energy/GDP position of some individual countries. As anticipated, Canada and Sweden fall in relation to the United States because of the hydroelectric factor. There is also some change in the relative position of the Netherlands and the United Kingdom owing to other characteristics of their transformation sectors (see chapter 6 for further details). This apart, Canada and the United States remain large consumers whether the measure of consumption is deliveries to final consumption sectors or total energy consumption.

Comparative Energy Consumption by Sector

So far we have been discussing aggregate energy/GDP ratios, but analysis on the aggregate level can mask significant differences in the composition of energy consumption at the sectoral level (see figure 9-1). Five consuming sectors are distinguished.

Household-commercial

The household-commercial sector includes energy consumed in households, in commerce, in governmental activities at all levels, in agriculture, and in miscellaneous other areas. It accounts for between 20 and 30 percent of total energy consumption.

Transport

The transport sector covers energy consumption by all forms of transport—air, road, rail, and inland waterway. Air and sea bunkers, that is, the fuel used to power international airplane flights and seagoing ships, have been separated out and treated as exports in order to limit this sector as much as possible to energy consumption in domestic transport, which enabled us to match GDP with total energy used in *domestic* economic activity. This sector accounts for about 10 to 22 percent of total energy consumption.

Industrial (including energy industries)

The industrial sector accounts typically for 30 to 35 percent of total energy consumed, with Japan being an exceptionally heavy consumer. This sector is divided into two parts. The first, which is by far the larger, consists of the energy consumed by nonenergy industries, of which the two largest are iron and steel, and chemicals.

The second consists of the energy industries' own consumption of energy. Such use includes the electricity consumed in the process of generating electricity, petroleum used in petroleum refining, and also electricity and gas distribution losses.

Transformation Losses

Those losses sustained in transforming energy from one form to another are included in this sector. The main component consists of the heat losses incurred in transforming coal and other fossil fuels into electricity. The sector accounts for approximately 15 to 20 percent of the total energy consumed in our countries. The transformation sectors of Canada and Sweden are somewhat higher than the other countries because of the inclusion of hypothetical heat losses following from the aggregation of hydroelectricity at fossil-fuel-equivalent efficiencies.

Nonenergy Use

The sector termed nonenergy use is relatively small, but sharply increasing. It entails the use of energy raw materials as feedstock for the petrochemical industry and other nonfuel and power applications, including grease, asphalt, road oil, and the like. It does not include natural gas, naphtha, or coal feedstocks. This sector varies most widely—from 2 to 17 percent of the total.

Contrasting Patterns of Energy Consumption

Our nine countries exhibit some broad similarity in energy consumption patterns stemming from the similarity in their economic structures and standards of living. But within the limits imposed by such similarities, there are marked contrasts in patterns of energy consumption. To identify these areas of differences at a more disaggregated level, table 3-2 gives sectoral energy/GDP ratios obtained by dividing the energy consumption of each sector by total GDP. The ratios for all sectors add up to the total energy/GDP ratio.

It must be emphasized that the sectoral ratios are used only as a first disaggregation of the overall energy/GDP ratio and as a point of departure for detailed sectoral analysis. They provide a means by which subsequent sector analysis can be linked additively to the energy/GDP ratios. They *do not* in themselves permit firm conclusions to be drawn about the comparative energy intensiveness of given sectors, either within a given country or between countries. Such comparisons require measures of sectoral output that are developed only in later chapters.

Comparisons of sectoral energy/GDP ratios within a country are particularly misleading. Thus, the small amount of energy consumed relative to GDP in the energy industry compared with the overall industry energy/GDP ratio completely obscures the fact that the energy industries are more energy intensive than industry as a whole. This discrepancy arises because the sectoral output measure of industry as a whole is very much larger than the sectoral output measure of the energy industry.

Comparisons of a given sectoral energy/GDP ratio between countries are less misleading because variations in sectoral output between countries of broadly similar economic structure are less marked than intersectoral variations. Nonetheless, there are differences between countries —indeed the analysis of these differences forms a major part of subsequent chapters—so that comparisons between countries though justified in a preliminary way must be treated only as a starting point for further analysis.

The first thing to notice in table 3-2 is that the ratios for the United States and Canada are higher for almost all sectors and in relation to almost all the countries. The higher U.S. overall energy/GDP ratio might have been the result of a mix of sectoral energy/GDP ratios, with those sectors that are higher than in other countries insufficiently offset by those sectors whose ratio is lower than elsewhere. But this is clearly not the case. The United States is consistently higher in almost every sector, and in those few cases where other sectors in other countries are somewhat higher than the United States, they are generally only slightly higher. There is then a pervasive tendency, reflected in all of the main consuming sectors, for U.S. consumption relative to GDP to be higher than that of other countries.

This higher consumption relative to GDP, then, affects all sectors, but it affects some much more than others. For example, consumption in the transport sector relative to GDP is particularly high in the United States and Canada compared with the other countries. This can be seen in part B of table 3-2 where consumption per unit of GDP is expressed as an

TABLE 3-2 *Energy Consumption by Sector, 1972*

Consumption by sector	U.S.	Canada	France	W. Germany	Italy	Netherlands	U.K.	Sweden	Japan
A. (tons oil equivalent per million dollars GDP)									
Total energy consumption	1,480	1,772	795	1,031	915	1,272	1,121	1,062	849
Transformation losses	250	401	140	170	133	164	254	267	147
Energy sector	135	128	57	74	48	100	81	33	48
Transport sector	327	305	117	132	136	134	145	121	105
Industry sector	309	388	219	299	282	254	318	275	330
Household-commercial	374	480	223	300	220	407	271	348	164
Nonenergy use [a]	86	70	38	55	96	213	53	18	54
All sectors minus transport	1,153	1,467	678	899	779	1,138	976	941	744
B. (index, United States = 100)									
Total energy consumption	100	119.7	53.7	69.7	61.8	85.9	75.7	71.8	57.4
Transformation losses	100	160.4	56.0	68.0	53.2	65.6	101.6	106.8	58.8
Energy sector	100	95.6	42.2	54.8	35.6	74.1	60.0	24.4	35.6
Transport sector	100	93.3	35.8	40.4	41.6	41.0	44.3	37.0	32.1
Industry sector	100	125.6	70.9	96.8	91.3	82.2	102.9	89.0	106.8
Household-commercial	100	128.3	59.6	80.2	58.8	108.8	72.5	93.3	43.9
Nonenergy use [a]	100	81.4	44.2	64.0	111.6	247.7	61.6	20.9	62.8
All sectors minus transport	100	127.2	58.8	78.0	67.6	98.7	84.7	81.6	64.5

Source: Data in appendix B.

[a] Energy raw materials used as feedstock for the petrochemical industry.

index relative to the United States. Taking U.S. consumption of energy in transport relative to GDP as equal to 100, all other countries (except Canada) are very much lower—between one-third and one-half of U.S. levels. In the other major sectors the differences between U.S. consumption and that of other countries is more mixed. In industry (excluding the energy industries) the variability is quite narrow. The unweighted average of industrial energy consumption relative to GDP in the Western European countries and Japan is only some 10 percent below U.S. levels. In the remaining sectors—household-commercial and transformation— some countries are very close to U.S. levels, although the variation between countries is somewhat greater than in the industrial sector. Consumption in these sectors relative to GDP averages some 25 percent below that of the United States.

The critical role of the transport sector in contributing to variations in consumption levels is illustrated in the last lines of parts A and B of table 3-2 where energy/GDP ratios for all sectors except transport are given. The exclusion of transport reduces the difference between U.S. and other-country levels of consumption noticeably. Excluding transport, consumption levels of non–North American countries rise to an unweighted average of 77 percent of U.S. consumption levels, compared with 67 percent if transport is included.

Table 3-3 presents the important role of transport in another way. Of the total differences between United States' and other countries' energy/ GDP levels, transport alone accounts for between 30 and 50 percent. In many countries, other sectors are also important in explaining total differences, but the one sector at this stage of disaggregation that is of major importance in all countries is transport.

With some modification the sectoral energy consumption data can be adjusted to give a different breakdown: between energy consumed directly by households in the form of gas, electricity, heating oil, and gasoline; and the remainder, which goes to intermediate uses before passing to final-demand categories in the form of energy embodied in a variety of goods and services. This is the form in which the input–output analysis in chapter 7 starts out. (See table 3-4.)

Following this classification, energy consumption associated with households' direct purchases of energy and power whether for heating, lighting, or gasoline, accounts for about 30 percent of the total. The U.S. and Canadian share is rather higher than the average European share. Japan, on the other hand, is markedly lower with only 18 percent of total

TABLE 3-3 *Difference Between U.S. Energy/GDP Ratios and Those of Other Countries, 1972*

Differences from U.S.	Canada	France	W. Germany	Italy	Netherlands	U.K.	Sweden	Japan
A. (tons oil equivalent per million dollars GDP)								
Total energy consumption	-292	685	449	565	208	359	418	631
Transformation losses	-151	110	80	117	86	-4	-17	103
Energy sector	6	78	61	87	35	54	102	87
Transport sector	22	210	195	191	193	182	206	222
Industry sector	-79	90	10	27	55	-9	34	-21
Household-commercial	106	151	74	154	-33	103	25	210
Nonenergy use [a]	16	48	31	-10	-127	33	68	32
B. (percentage of total energy)								
Total energy consumption differences	100.0	100.0	100.0	100.0	100.0	100.0	100.0	100.0
Transformation losses	51.7	16.1	17.8	20.7	41.3	-1.1	-4.1	16.3
Energy sector	-2.1	11.4	13.6	15.4	16.8	15.0	24.4	13.8
Transport sector	-7.5	30.7	43.4	33.8	92.8	50.7	49.3	35.2
Industry sector	27.1	13.1	2.2	4.8	26.4	-2.6	8.1	-3.3
Household-commercial	36.3	22.0	16.5	27.3	-15.9	28.7	6.0	33.3
Nonenergy use [a]	-5.5	7.0	6.9	-1.8	-61.1	9.1	16.3	5.1

Notes: In part A, a negative number indicates that U.S. energy/GDP ratio is lower than that of other country. In part B, 100 refers to the total difference between United States and other countries. In the case of Canada, the difference represents a higher level of consumption; in all other countries, lower.

Source: Data in appendix B.

[a] Energy raw materials used as feedstock for the petrochemical industry.

TABLE 3-4 *Differences Between United States and Other Countries in the Direct Household and Other Sectors of Total Final Energy Consumption/GDP Ratios, 1972*

Differences from U.S.	Canada	France	W. Germany	Italy	Netherlands	U.K.	Sweden	Japan
	tons oil equivalent per million dollars GDP							
Total final energy consumption	−240	721	450	565	233	408	416	608
Direct household	−72	270	207	222	128	155	117	336
utilities	−109	139	87	126	6	30	4	176
gasoline	36	131	121	96	122	125	113	159
Other	−169	450	242	342	104	252	298	271
	percentage							
Total final energy consumption	100.0	100.0	100.0	100.0	100.0	100.0	100.0	100.0
Direct household	30.0	37.4	46.0	39.3	54.9	38.0	28.7	55.3
utilities	45.0	19.2	19.2	22.0	2.6	7.4	0.9	29.0
gasoline	−15.0	18.2	26.9	17.0	52.4	30.6	27.2	26.2
Other	70.0	62.4	53.8	60.5	44.6	61.8	71.6	44.6

Notes: Total final energy consumption includes heat losses. Negative numbers indicate higher total final energy consumption ratio than that of the United States.
Source: Data in appendix **B.**

energy consumption devoted to direct energy purchases by households. Such a difference is wide enough to swamp the deficiencies of the data and indicates a significant contrast in patterns of energy consumption between Japan and other countries.

Although direct purchases of energy by households account for only 30 percent of total energy consumption, they account for considerably more—45 percent or more—of the overall difference between the U.S. on the one hand, and Western European and Japanese energy/GDP ratios on the other. And of this difference, gasoline consumption accounts for about 30 percent and utilities for 15 percent. Again, this presentation of data draws attention not only to the importance of the transport sector in explaining higher U.S. energy consumption relative to GDP compared with other countries, but to the importance of passenger transport in particular.

Country Profiles

In the preceding discussion, emphasis was on sectoral differences among countries, with some reference to major differences between geographical groupings of countries, North America, Western Europe, and Japan. Although much of the analysis in this study will be made in terms of regional groupings because this is where the major differences lie and because the quality of our data in some sectors will not support detailed intercountry analysis, there are, nevertheless, significant differences in energy consumption between countries within the same geographical grouping—between the United States and Canada, for example, and between France and Italy. Even if it is not possible in a study of this nature[2] to analyze such differences in detail, some awareness of them is useful background for the interpretation of later results.

United States, together with Canada, is the most energy intensive of the nine countries. High energy consumption relative to income is reflected in all sectors, particularly in transport. By comparison with most countries the United States enjoys a high degree of self-sufficiency in energy supplies, of which about 80 percent are oil and gas.

[2] Detailed studies comparing the energy consumption of one country with that of another include: A. Doernberg, *Comparative Analysis of Energy Use in Sweden and the United States* (Upton, N.Y., Brookhaven National Laboratory, 1975); Lee Schipper and A. J. Lichtenberg, *Efficient Energy Use and Well-being: The Swedish Example* (Berkeley, Calif., Lawrence Berkeley Laboratory, University of California, 1976); Stanford Research Institute, *Comparison of Energy Consumption Between West Germany and the United States* (Menlo Park, Calif., SRI, 1975).

Canada's level and pattern of energy consumption are very similar to those of the United States, although the industrial and household-commercial sectors use somewhat more energy relative to GDP. Of all the countries, Canada has the most diversified fuel mix and, in 1972, was alone in being completely self-sufficient—in fact, a net exporter.

France consumes the least energy relative to output of all our countries, using about one-half of the energy consumed per unit of output that the North American countries do. France consumes less per unit of GDP in all sectors but very much less in transport. France has a very high dependence on imported petroleum.

Italy consumes at North American levels in the industrial and non-energy sectors, owing to large refinery and petrochemical operations serving the export trade as well as the domestic market. As with other European countries the greatest sectoral contrast with United States is in transport. Even for a European country, Italy has an exceptionally high degree of energy import dependence.

Netherlands uses by far the most energy relative to GDP of the European countries, only 13 percent below U.S. levels. Like Italy, the Netherlands has a large refinery and petrochemical complex serving neighboring countries as well as themselves. But unlike Italy, consumption in the household-commercial sector is also high. Despite the necessity of importing petroleum for petroleum-based industries, the Netherlands has a relatively high degree of self-sufficiency because of the exploitation of domestic natural gas. In fact, it is the only European country to be consuming domestically produced natural gas on a large scale.

The United Kingdom is also a relatively high European user. Consumption of energy relative to GDP in both the industrial and transformation sectors is similar to that in the United States, but, as with other European countries, transport is much lower. Because the United Kingdom has large domestic supplies of coal, the 1972 degree of dependence on imported fuels was lower than in many countries.

Sweden has high levels of consumption in both the household-commercial and industrial sectors, but again, like other European countries, very much lower in transport. Sweden has an exceptionally high energy import dependence.

Japan, overall, is one of the least energy-intensive countries, similar to France in total but very different in sectoral composition. As in several European countries, energy consumption relative to GDP in the industrial sector is very similar to U.S. levels. Consumption in the transport and household-commercial sectors, however, is exceptionally low, even by European standards. Japan also has a high degree of import dependence.

Summary

In sum, it can be said that the North American countries consume about 50 percent more energy relative to GDP than the countries of Western Europe and Japan. Put another way, the average energy consumption relative to GDP of Western Europe and Japan is some 30 percent below U.S. levels.

The higher U.S. energy consumption relative to output is pervasive: it is reflected in almost all sectors, though more in some than others. Of the four major sectors—industrial, transformation losses, household-commercial, and transport—there was least variability in the industrial sector, where the unweighted average of energy consumption relative to GDP in the Western European countries and Japan was only some 10 percent below U.S. levels. In both the household-commercial and transformation loss sectors, the European countries and Japan show more variability, averaging consumption some 25 percent below that of the United States. But by far the most variability is shown in the transport sector where European and Japanese consumption is on the average 60 percent lower than in the United States. In the household-commercial, industrial, and transformation sectors there is considerable variation in consumption levels between the Western European countries and Japan, but in transport there is a striking similarity in experience. All of the seven countries have very similar transport energy consumption relative to output.

Similarities in sectoral energy/GDP ratios as they appear in table 3-4 can, however, obscure major contrasts in the mix and energy intensity of activities carried out inside the sectoral definition. It is not possible, for example, to conclude that specific energy intensity within the industrial sector is higher in Japan than in the United States on the basis of sectoral energy/GDP ratios alone. Further disaggregation distinguishing between structural and intensity characteristics is necessary. This will be done by sector in chapters 4, 5, and 6.

PART II

Sectoral Approach

CHAPTER 4

Household-commercial Sector

THE PURPOSE of this chapter is to demonstrate the extent to which differences in household and commercial use of energy contribute to differences in the overall energy/GDP ratios between the United States and other countries. A brief introduction surveys energy use in the sector as a whole. The next section is devoted to the major topic of the chapter: an analysis of differences between the United States and the other countries in energy use in household space conditioning. Then in the interest of comprehensive coverage, summary analyses of the energy consumed in other household, commercial, and agricultural uses are also made. Estimating procedures for this chapter are discussed in appendix C.

Although for convenience we call this sector the "household-commercial" sector, its more accurate title is the one used by OECD, which is the "other sector"—that is, the sector that treats all energy consumption not included within the more clearly defined sectors. As such, this sector includes energy consumption in a wide variety of areas such as households, office buildings, street lighting, agriculture, government operations on both the national and local levels, some military uses,[1] shops, restaurants, and miscellaneous handicrafts. Data problems compelled us to restrict our analysis of energy consumption in this sector to the threefold grouping of household, commercial, and agricultural. Of these three categories, household is for all countries by far the largest—about 60 percent of the total.[2] The commercial category accounts for about 30

[1] The distribution of "military uses" in the energy consumption statistics is unclear. Part of military uses may be included in the industrial sector or in stock changes. Military uses in the United States account for about 1.4 percent of total energy consumption and if all of them were included in the "other" sector, for about 5 percent of that sector.

[2] Household use in this study is considered to be the consumption of fuel and power by households for lighting, heating, cooking, and the like. It does not include gasoline consumption, which, in the sectoral analysis is included under transport. Thus, household consumption is not the same as the direct energy purchases by households referred to in chapters 7 and 8 that include energy consumption of gasoline and also the heat losses associated with electricity consumption.

percent, and agricultural for a relatively small share (typically under 10 percent) of the household-commercial sector consumption (see table C-1 in appendix C). Compared with the strikingly different patterns of energy consumption in the transport sector between North America and Europe, there is relatively little variation in these shares between countries.

Another difference between this sector and the transport sector is that relative to GDP, the United States is not invariably the largest consumer. The Netherlands and Canada are larger consumers in the household-commercial sector (see table 4-1). The other European countries are much lower than the United States, which implies considerable variation within Europe. Consequently, it is much more difficult to contrast European experience with that of the United States.

Differences in energy consumption in household-commercial uses relative to GDP do, however, account for a substantial if varying part of the total difference in energy/GDP ratios between the United States and the other countries, and within the household-commercial sector, the household category alone accounts for between 10 and 20 percent of the variation in total energy/GDP ratios between many European countries and the United States.

Household Category

Energy consumption in the household category includes a wide variety of uses—heating, cooling, lighting, cooking, refrigeration, and the energy involved in using domestic appliances. Because of data limitations, we have based our analysis on two main functional groupings within the household category—energy consumption for (1) heating and cooling (which we shall term space conditioning) and (2) non-space conditioning purposes. For the purposes of analysis, a sharp distinction is drawn between these two uses of energy within the household, and a note of caution is in order. The whole concept of space conditioning, in contrast to other energy uses within the household, is in some respects contrived. A large family, spending much time in the kitchen or in a room adjacent to cooking facilities, may not need space heating as such; heat generated by people (each person represents a source of 100 watts) and the heat and steam from the kitchen may be more than sufficient to ensure adequate heat. Often it is difficult conceptually to assign a part of total energy consumption to purely space heating as opposed to other purposes.

TABLE 4-1 Energy Consumption in the Household-commercial Sector Relative to GDP, 1972
(tons oil equivalent per million dollars GDP)

Energy consumption	U.S.	Canada	France	W. Germany	Italy	Netherlands	U.K.	Sweden	Japan
Consumption of sector	374	480	223	300	220	407	271	348	164
Difference from U.S.	—	−106	151	74	154	−34	103	26	210
Household category	218	289	126	164	134	245	180	231	85
Difference from U.S.	—	−71	92	54	84	−27	38	−13	133
Commercial category[a]	132	142	84	129	72	120	80	106	72
Difference from U.S.	—	−10	48	3	60	12	52	26	60
Agriculture category	25	48	13	8	14	43	11	12	7
Difference from U.S.	—	−23	12	17	11	−18	14	13	18

Source: Data in appendix B.
[a] The commercial category includes miscellaneous.

In poorer communities particularly, the division between heating and nonheating uses is more likely to be arbitrary.

Space Conditioning

Energy consumption for space conditioning is by far the larger of the two uses, accounting for 70 to 80 percent of total energy consumed in the sector and for most of the total variation in household-commercial sector energy consumption relative to GDP. Within space conditioning, heating is the most important as far as energy consumption is concerned. In Europe in 1972, air conditioning was assumed to be negligible, and even in the United States, air conditioning in household and commercial use combined accounted for only 5 percent of total energy consumption in those categories.

As table 4-2 shows, there is considerable variation in the consumption of space conditioning energy relative to GDP among countries. Canada, Sweden, and the Netherlands consume substantially more energy relative to GDP for space conditioning than the United States, and the others rather less. Japan is by far the lowest, consuming only one-third as much heating and cooling energy relative to GDP as the United States. As noted in chapter 2, the Swedish data are believed not to include district heating, which accounts for some 20 percent of the total. Consequently, the Swedish consumption may be underestimated relative to other countries.[3]

The amount of energy consumed for space conditioning depends on a number of factors: the outside temperature in both summer and winter, the characteristics of the housing stock, the amount and effectiveness of insulation, the efficiency of heating systems, and differing heating habits. Here we shall try to analyze the effect of differences in climate, housing stock, insulation, and heating systems in order to isolate how much of the observable difference is due to what can conveniently be termed "different heating habits." This would include such things as starting and stopping heat at different temperatures, the heating of unoccupied areas, and heating to higher temperatures.

---

[3] Lee Schipper, whose study of U.S. and Swedish energy consumption permitted more detailed treatment of energy consumption in Sweden than the present study covering nine countries, disagrees. He considers our data on residential consumption high enough to include district heat. See: Lee Schipper and A. J. Lichtenberg, *Efficient Energy Use and Well-being: The Swedish Example* (Berkeley, Calif., Lawrence Berkeley Laboratory, University of California, 1976).

TABLE 4-2 *Energy Consumption in the Household Category, 1972*

Energy consumption	U.S.	Canada	France	W. Germany	Italy	Netherlands	U.K.	Sweden	Japan
	million tons oil equivalent								
Household category	256.9	29.9	27.2	40.3	19.0	12.0	34.3	9.4	30.7
Space conditioning	194.3	24.7	21.8	31.3	15.2	10.3	25.8	8.3	17.0
Non-space conditioning	62.6	5.2	5.4	9.0	3.8	1.7	8.5	1.1	13.7
	tons oil equivalent per million dollars GDP								
Household category	218	289	126	164	134	245	180	231	85
Space conditioning	165	239	101	127	107	210	136	204	47
Non-space conditioning	53	50	25	37	27	35	45	27	38
	tons oil equivalent per million dollars GDP								
Household category, difference from U.S.	—	-71	92	54	84	-27	38	-13	133
Space conditioning	—	-74	64	38	58	-45	29	-39	118
Non-space conditioning	—	3	28	16	26	18	8	26	15

Source: Data in appendix B.
Note: Blanks = not applicable.

CLIMATE. First, how much of this difference in energy consumption for space conditioning can be attributed to differences in climate? The nine countries have a wide variety of climate, ranging from arctic to semi-tropical, so that climate would normally be expected to have a major influence on the amount of energy consumed in heating and cooling. The method chosen for assessing the influence of climate on space conditioning fuel consumption was to calculate how much fuel other countries would consume for space conditioning if they had the same climate as the United States. This was done in two stages.

First, an adjustment for the cooling element of space conditioning was made. Since in 1972 the United States was the only significant user of energy for household cooling among our countries, the adjustment for the assumed hotter American summer was made by subtracting from the U.S. total the amount of energy used in household air conditioning that year (about 7 million tons oil equivalent, or some 3 percent of total household consumption).

There was some difficulty in treating consumption of energy in air conditioning as a purely climatic adjustment. The United States has a much larger part of its territory and population in continental and semi-tropical areas than the other countries (with the possible exception of Italy), which provides some justification for the procedure. However, not all of this consumption is necessarily related to summer temperatures higher than those in Europe, but represents a standard of comfort additional to strictly climatic needs. The deduction of total energy consumed in air conditioning as a climatic adjustment would therefore overestimate to some extent the influence of hot climate on consumption of space conditioning energy in the United States.

It would assign to "climatic differences" some part of space conditioning energy that should more properly be assigned to differences in heating habits. But whichever way air conditioning fuel is treated, it does not account for a significant part of the difference in energy consumption between the United States and other countries, because of the relatively small quantities involved.

The second step was to adjust for differences in winter temperatures. This was done by standardizing the heating fuel consumption of all countries by assuming that all countries have the same winter temperature, expressed in "degree days," as the United States.[4]

---

[4] Degree days measure the difference between actual temperature and the temperature at which heating is needed (the threshold temperature). If the average temperature in a single 24-hour period in a given locality is 55°F, and the threshold tem-

The results of these climatic adjustments for both heat and cold are given in tables 4-3 and 4-4. Line (1) of table 4-4 gives the total unadjusted difference between energy consumption for residential space conditioning relative to GDP in all countries compared with the United States. A positive number means that the U.S. consumes more energy in space conditioning than the other country. This difference is initially narrowed by the amount of energy consumed in air conditioning in the United States [line (2)]. But in almost all cases, due to the relatively small number of U.S. heating degree days, the original difference is widened by the standardization of energy consumption for heating [line (3)]. As the heating adjustment is quantitatively much greater than the cooling adjustment, the net effect is to increase the difference between the United States and other countries. That is to say, after corrections for climatic differences, the United States is found, relative to the other countries, to be a larger consumer of energy for space conditioning relative to GDP than it was before correction.

For some of the countries—France, Germany, United States, Japan—the adjustment for climate does not make a lot of difference, but for Canada and Sweden, which started out with higher heating energy consumption relative to GDP than the United States, it makes a considerable difference. In these countries, the climate explains the entire difference, and more than the difference, in actual consumption.

The effect of climatic correction is therefore to widen the gap between the United States and other countries. The former emerges as the largest consumer of residential space conditioning energy relative to GDP. Whatever causes the higher U.S. consumption of energy for space conditioning relative to GDP, climate is not responsible, and it is necessary to examine other factors such as the nature of the housing stock.

HOUSING STOCK. The effect of different housing characteristics—size and type of house—clearly influences consumption of fuels for heating purposes. In carefully controlled experiments it is easy to study the effects of larger house size, better insulation, and the like, on energy

---

perature is 65°F, then the number of degree days during that period would be 10. The same circumstances on two days would yield 20 degree days. In such a way, degree days are added cumulatively over the winter months to give a total for the whole heating season. A country average can be calculated by weighting degree days by population of the different localities. (See appendix C for details of degree day estimations.) Note that degree days include only positive sums. An average temperature above the threshold temperature means that heating is not needed. The European threshold of 16°C or 60.8°F was used.

TABLE 4-3 *Energy Consumption for Household Space Conditioning, Corrected for Climate, 1972*

Energy consumption	U.S.	Canada	France	W. Germany	Italy	Netherlands	U.K.	Sweden	Japan
For space conditioning (tons oil equiv. per $ million GDP)									
Cooling	165	239	101	127	107	210	136	204	47
Heating	6	—	—	—	—	—	—	—	—
	159	239	101	127	107	210	136	204	47
Degree days (centigrade at 16° threshold)	2,030	3,700	2,200	2,600	1,700	2,725	2,200	3,800	2,000
For heating adjusted for climatic differences (tons oil equiv. per $ million GDP)	159	131	93	99	136	156	125	109	48

Source: Data in appendix B.
Note: Blanks = not applicable.

TABLE 4-4 Differences Between United States and Other Countries in Energy Consumption for Household Space Conditioning, Corrected for Climate, 1972

(tons oil equivalent per million dollars GDP)

Differences from U.S.	Canada	France	W. Germany	Italy	Netherlands	U.K.	Sweden	Japan
(1) Total difference	−74	64	38	58	−45	29	−39	118
(2) Cooling	6	6	6	6	6	6	6	6
(3) Heating adjustment	−108	−8	−28	29	−54	−11	−95	1
(4) Total climatic adjustment	−102	−2	−22	35	−48	−5	89	7
(5) Total difference after climatic adjustment	28	66	60	23	3	34	50	111

Note: A positive number means the United States consumes more energy in space conditioning than the other country.

consumption; but it is difficult to generalize from controlled experiments about stocks of houses in various countries. Each country has a mix of large and small houses, some adequately insulated in relation to the outside environment and some not. The range of climatic variations in some countries, particularly the United States, makes a judgment about "average" levels of insulation highly suspect.

The question was to see whether there is any indication of a *systematic* difference in housing size and characteristics associated with heat losses among countries which might contribute to explaining variations in heating fuel consumption after climatic corrections have been made. Reliable and comparable data on these matters were not available; it was necessary to rely on indirect indications given by data on housing characteristics from United Nations sources and the *ICP Report*. These data are in table 4-5.

The United States and Canada are commonly supposed to have larger houses than most countries, and data available give some support to this view. But other countries also have large houses, in particular, Sweden and the United Kingdom, so that size alone may not account for the larger U.S. consumption.

Heat losses occur through walls, roofs, and windows. The larger the outside wall surface and the greater the number of windows, the more energy will be needed for heating. A country with a stock of housing consisting largely of single family houses, other things being equal, should have a greater heat loss. Such a country would require more heating fuel than a country whose housing stock has a large proportion of apartments. Note, however, that the smaller heat loss of apartments deriving from the smaller outside wall area and fewer windows may be compensated for, as appears to be the case in Sweden, by higher energy consumption on the part of tenants whose utility costs are included in rent and therefore have no incentive to economize on heating bills, and by difficulties in controlling temperatures in apartments.

From the fragmentary data available, it appears that the United States, Canada, United Kingdom, and the Netherlands have a high proportion (about 70 percent) of single family houses, while housing in France, Germany, Sweden, and Italy is about equally divided between single family and multiple family houses. From the point of view of heat losses, however, the United Kingdom and Netherlands differ from the United States and Canada in that their single family units are more likely to consist of row-type houses with only two outside walls and proportionately fewer windows than the North American type with four outside walls.

TABLE 4-5 *Characteristics of Dwellings, 1972* [a]

Characteristics	U.S.	Canada	France	W. Germany	Italy	Netherlands	U.K.	Sweden	Japan
Number of dwellings (thousands)	68,699	6,259	18,120	22,852	17,434	4,000	19,167	3,181	25,591
Percentage single family	69	70	52	50	60	67	70	43	—
Percentage multifamily	31	30	48	50	40	33	30	57	—
Percentage owner-occupied	76	65	43	34	51	36	50	35	49
Average floor space per dwelling (square meters)	95.6	—	52.4	62.3	54.4	—	84.5	(60.0)	—
Index of housing characteristics	100	100	73	73	69	74	80	83	69

Notes: Blanks = not available. Figure in parentheses is RFF estimate.
Sources: United Nations, *Statistical Yearbook, 1974* (New York, UN, 1973); United Nations, *Yearbook of Construction Statistics, 1964–73* (New York, UN, 1975); *ICP Report.*
[a] Or nearest year.

Thus, in the scale of heat losses, the United Kingdom and Netherlands are likely to occupy a middle ground between North America and the other countries.

For ease of handling, these two characteristics—size and proportion of single family dwellings in the total—have been combined, in a somewhat qualitative manner, into a single index. If the housing characteristics of the United States (and Canada) equal 100, most of the European countries and Japan have a value of 70, reflecting smaller houses and, in most cases, a higher proportion of multifamily units. The United Kingdom, owing to a larger proportion of single family residences, and Sweden, with a larger size of house, are somewhat higher (80).

This index is applied in table 4-6 to climate-corrected data to illustrate how much more energy relative to GDP would be needed for heating by other countries if they had housing similar in terms of size and single–multiple unit mix to the United States. Because all other countries except Canada have smaller houses on the average and because most have a larger proportion of multifamily units, they would all need to consume more fuel for heating relative to GDP, if they had the U.S. size and mix. Conversely, if U.S. housing characteristics matched those abroad, U.S. heating energy consumption relative to GDP would be expected to be lower. Part of the total difference in consumption of heating fuels (per million dollars of GDP) between United States and other countries is due therefore to the characteristics of U.S. houses, which typically are larger and more frequently of the single family type. For many countries, housing differences account for about one-half of the total, leaving the other half of the difference to be accounted for by a variety of factors such as different insulation standards, efficiency of the heating systems, heating habits, and the like.

Before considering these other factors further, two observations should be made. The first is that the characteristics of the housing stock in both Italy and the Netherlands *overexplain* the total differences between these countries and the United States, leaving a substantial negative sum—that is, a higher energy use relative to GDP than the United States—to be explained. For Italy, the difficulty may be in the original estimates of heating fuel, or in the degree-day data, both of which are particularly susceptible to error. The Netherlands is different. Its high consumption for heating purposes relative to GDP appears to be well established, implying a high intensiveness of energy use in household heating.

The second observation is that the partition of the total difference in space conditioning energy consumption between housing character-

TABLE 4-6 *Differences in Energy Consumption Between United States and Other Countries Accounted for by Housing Characteristics and Heating Habits, 1972*
(absolute figures in tons oil equivalent per million dollars GDP)

Item	U.S.	Canada	France	W. Germany	Italy	Netherlands	U.K.	Sweden	Japan
Energy consumption for space conditioning, adjusted for climate differences	159	131	93	99	136	156	125	109	48
Total difference in space conditioning after climatic adjustment	—	28	66	60	23	3	34	50	111
Index of housing characteristics (U.S. = 100)	100	100	73	73	69	74	80	83	69
Energy consumption for space conditioning, adjusted for climate difference and standardized for housing stock	159	131	127	136	197	211	156	131	70
Difference due to housing characteristics	—	0	34	37	61	61	31	22	22
Difference due to unexplained residual (attributed to heating habits)	—	28	32	23	-38	-52	3	28	89

Source: Data in appendix B.
Note: Blanks = not applicable.

istics and heating habits is highly sensitive to relatively modest movements in our rough "housing index." A fall by, say, 10 to 20 percent
in the index for most countries would result in the *total* difference in
energy consumption relative to GDP being explained by differing housing. Given the uncertainty of the basic data, such a possibility should
not be ruled out. On the other hand, the average size of a house, upon
which the index is based, may underestimate the total amount of housing
space in other countries relative to the United States. In this case, the
role of housing in the final difference would be diminished. On balance,
however, it seems reasonable to conclude that the larger, more typically
single family dwelling of the United States is responsible for a substantial
part of that country's higher consumption of energy for heating per unit
of GDP, relative to Europe and Japan.

INSULATION. Part of the difference remaining after climatic adjustment
and after taking into account differing housing characteristics (the unexplained residual of table 4-6) may be due to differences in insulation practices. Good insulation can frequently reduce heat losses substantially,
so the obvious question must be asked: is there any reason to believe
that standards of insulation vary markedly between countries? Among
our countries, Sweden is conspicuously better insulated than the others.[5]
U-Values (which measure rate of energy loss across area per unit of temperature difference between the inside and outside surfaces) appear to
be typically half as high as in the United States, Netherlands, and the
United Kingdom. Canada, too, appears to approach Swedish levels of
heating efficiency in the coldest provinces. This could mean that most,
perhaps all, of the difference between the U.S.–Swedish residual differences (table 4-6) could be due to improved insulation, and this could
account for a good part of the Canadian difference, too. (It should be
recalled that the original difference between the United States and Sweden is probably overestimated if district heating is excluded.)

Generally high standards of insulation are likely to be restricted to
Canada and Sweden, with their long and predictably severe winters. These
countries have, in addition, greater economic incentive. No matter how
low heating fuel prices are—and they have been relatively low in both
Sweden and Canada—heating bills in such climates will be very high,
offering householders strong incentive to install tight insulation.

[5] Schipper and Lichtenberg, *Efficient Energy;* J. A. Over, ed., *Energy Conserva-
tion: Ways and Means* (The Hague, Future Shape of Technology Foundation, 1974;
National Development Office, *Energy Conservation in the United Kingdom* (London,
HMSO, 1974).

Apart from these two countries, there is probably little variation in insulation standards among countries. United States insulation standards may, on the average, be somewhat better than in other countries, except for Canada and Sweden, as the United States and Canada have a particularly high proportion of owner-occupied dwellings. This could provide incentives to improve insulation, since the benefits in the form of lower fuel requirements are gained by those who incur the cost of insulation. In countries with a lower proportion of owner-occupied dwellings, incentives to improve insulation may be weaker.

HEATING SYSTEMS. The remaining difference in heating energy consumed relative to GDP is due to a mixture of reasons which it is probably not useful to partition formally, but which nonetheless should be discussed. First is the efficiency of the heating system, defined in this case as the amount of useful energy obtained from a given input of energy. Coal burned in an open fireplace, for example, will require much more fuel to provide a given room temperature than a furnace fired by natural gas. In addition, the efficiency of the equipment itself may also vary: it is said that the efficiency of an oil furnace can be raised from 50 to 70 percent by good maintenance. In these ways, the efficiency of the heating system can vary considerably, depending on the type of installation, the fuel used, and maintenance standards.

In the case of central heating systems, electric heat, which has no chimney loss, is the *most efficient* (though note at this point we are not incorporating the waste heat involved in the production of that electricity). Electricity at 94 percent efficiency is far above that of gas and oil central heating, which at first sight, surprisingly, are not very much more efficient than coal-fired central heating (see table 4-7).

The variation in efficiency between coal and the other fuels is much wider in noncentral heating systems, where coal burned in open fires, falls to 30 percent—half as efficient as oil and gas and only one-third as efficient as electricity.[6] If the energy losses sustained in the production and delivery of these fuels to the final consumer are included, these efficiencies change radically; electricity, bearing its share of heat losses, now becomes less efficient than coal and only half as efficient as oil and gas. Given the structure of our analysis, we must analyze efficiencies

---

[6] This is now thought to be an underestimate, at least for the United Kingdom, which is the only country of our nine that still relies significantly on coal burned in open fires for space heating. The U.K. National Coal Board now shows efficiencies of nearer 45 to 50 percent when heat gained from the warm flue stack runs up through two floors.

TABLE 4-7 *Efficiency of Heating Systems, 1972*

Item	U.S.	Canada	France	W. Germany	Italy	Netherlands	U.K.	Sweden	Japan
Consumption of heating fuel (million tons oil equiv.)	194.3	24.7	21.8	31.3	15.2	10.3	25.8	8.3	17.0
Coal	—	—	3.5	2.3	0.5	0.2	12.5	—	—
Gas	89.6	6.5	1.2	1.6	1.6	5.8	7.2	—	—
Electricity	22.4	2.0	—	2.0	—	—	2.5	0.4	—
Petroleum products	82.2	16.2	17.0	25.4	13.2	4.2	3.5	7.9	17.0
Central heating penetration (U.S. = 100)	100	90	35	50	25	32	30	91	n.a.
Heating efficiency [a] (percentage)									
Coal	—	—	30	30	30	30	30	—	—
Gas	70	70	63	63	63	70	63	—	—
Electricity	94	94	—	99	—	—	99	94	—
Petroleum products	70	70	66	66	63	63	70	70	63
Useful energy (million tons oil equiv.)	141.3	17.8	13.1	20.5	9.5	6.9	13.3	5.9	10.7
Coal	—	—	1.1	0.7	0.2	0.1	3.8	—	—
Gas	62.7	4.6	0.8	1.0	1.0	4.1	4.5	—	—
Electricity	21.1	1.9	—	2.0	—	—	2.5	0.4	—
Petroleum products	57.5	11.3	11.2	16.8	8.3	2.7	2.5	5.5	10.7
Useful energy as percentage of total consumption of heating fuel	73	72	60	65	63	67	52	71	63

Notes: Blanks = not applicable. n.a. = not available.

[a] Based on data in National Economic Development Office, *Energy Conservation in the U.K.* (London, HMSO, 1974); American Institute of Physics, "Efficient Use of Energy," *Proceedings of AIP Conference*, no. 25 (New York, 1975); and North Atlantic Treaty Organization, *Technology of Efficient Energy Utilization*, Report of a NATO Science Committee conference held at Los Ares, October 1973.

disregarding losses in transformation, which are dealt with in chapter 6. Since electric heating is not widely used in most of our countries, the main conclusion of this section holds whether electric heating is assigned the low or the high efficiency.

These efficiencies, adjusted for estimated central heating penetration, were applied to each country's fuel mix to assess the influence of fuel mix and heating equipment—though not equipment maintenance—on variations in heating fuel consumption. The effect is to widen the disparity between the United States and other countries; that is, insofar as the useful heat delivered depends on the degree of central heating penetration and the absence of coal, the United States manages to achieve fewer losses than other countries and therefore derives a larger amount of useful energy out of a given input (see table 4-7). The sharpest divergence among our countries is with the United Kingdom, where a substantial part of fuel used for space heating consists of coal burned in open grates. This means that the useful energy delivered from a given unit is much lower than in other countries.

In terms of table 4-6, the higher efficiency of the U.S. heating system would increase the difference between United States and other-country consumption, leaving a larger residual to explain. Consequently, much of the residual, "unexplained" difference between U.S. heating energy consumption relative to GDP and that of the other countries must be due to a higher intensiveness of space conditioning consumption (that is, more energy-using heating habits) relative to GDP, above and beyond climate, housing characteristics, and efficiency of heating systems. This could take the form of starting to heat at a higher outside temperature—indeed this practice is institutionalized in the higher (65°F) threshold in degree day calculations in North America. Other habits also have an influence. Indoor temperatures may be kept higher in the United States than in other countries, though both Swedish and German indoor temperatures are reported to be the same as those in the United States.[7] Other countries place greater reliance on supplementary means of keeping warm (heavier clothing). Some of the difference in heating habits between countries may also be incorporated in the climatic variables. It has already been suggested that the cooling allowance made for the United States incorporates some element of a higher standard of comfort. Similarly, allowance for bigger houses in the United States contains

[7] Schipper and Lichtenberg, *Efficient Energy;* Stanford Research Institute, *Comparison of Energy Consumption Between West Germany and the United States* (Menlo Park, Calif., SRI, 1975).

an element of difference in heating habits because larger housing involves an increase in unused heated areas.

In summary, the purpose of this section has been to analyze differences in energy consumption for space conditioning relative to GDP between the United States and other countries. The analysis shows the United States, always on the high side, to be the largest consumer after adjustment for climatic differences. This is because the United States has a warmer average climate than the other countries, except for Italy, so that the relatively small adjustment to represent cooling needs is swamped by the larger heating needs of other countries. Perhaps one-half of this enlarged difference between the space conditioning consumption, relative to GDP, of the United States and other countries may be due to the prevalence of the larger, single family house. Since the U.S. heating system is relatively efficient, the other half is assumed to be caused by differences in heating habits.

For the United States, therefore, two factors—climate and the efficiency of the heating plant—tend to reduce heating energy consumption compared with other countries. But these two factors are more than offset by the consumption associated with larger houses (of which a larger proportion are single family houses) and different heating habits.

Unlike the transport sector where a generally similar European experience contrasts sharply with the U.S., fuel consumption for heating between countries within Europe varies considerably. Rather than compare the U.S. with a Western European average, it is more useful to recall from our discussion three main conclusions applying to other countries to illustrate some of the different facets of heating energy use.

1. Climate correction reduces Canadian and Swedish consumption to the level of other countries. Swedish consumption may, however, be underestimated due to exclusion of district heating. The tight insulation induced by the severe winters in these countries results in their lower energy/GDP relationship (after correction for climate and housing stock) compared with the United States, and even with the other European countries. This suggests that there is a critical number of degree days above which tight insulation becomes prevalent, probably because of high heating costs, and that such insulation more than compensates for climatic conditions, reducing the energy intensiveness to lower levels than in less severe climates. Within Canada, for example, those provinces with particularly severe climates consume less heating energy per degree day than do those with a lower number of degree days.

2. Within Europe, the largest consumers after climate correction are Italy, Netherlands, and United Kingdom. The high U.K. consumption appears to be explained by the inefficiency of the heating plant, that is, the greater reliance on coal and briquettes burned in an open grate. We have been unable so far to throw much light on the high consumption in Italy and the Netherlands. While part of this may be due to data deficiencies—Italy may be particularly prone to error in both heating fuel estimates and degree day data—it does appear that the Netherlands has a particularly high consumption of heating energy relative to GDP, perhaps associated with cheap natural gas.

3. Finally, Japanese consumption is very low, substantially lower than even the lowest European country. This cannot be due to any error or bias in estimating procedures. Japan's total sectoral consumption starts out low, and no matter how that total is distributed, consumption remains low.

The reasons for low Japanese consumption are significant differences in housing design and conditions of a kind which are not easy to capture by our index of housing characteristics, especially as they affect heating fuel consumption. Houses are typically built of wood and paper designed to minimize the discomforts of the hot wet summer. Many houses do not have bathrooms, with people relying instead on public baths. A large number of young people, estimated at 3 million, live in dormitory accomodations. Rooms are used in a different way, with a single room being used by all family members for dining, watching television, and entertaining friends. Such housing conditions tend to be very economical in the use of heating fuels. Until recently the charcoal heater, which is no doubt still used in some parts of the country, and which burns a noncommercial fuel not included in our data, was the principal facility for heating. It has now been replaced by the small, portable kerosine heater.

Such housing characterized by modest fuel consumption is undoubtedly associated with the much smaller proportion of private consumption expenditure in GDP in Japan as compared with any of the other countries. If energy consumption in the household category is expressed relative to consumption expenditure rather than GDP, the difference between Japan and the others narrows considerably. Even so, Japan still remains well below the others.

Non-space Conditioning

This category consists of the amount of energy used in operating home equipment and appliances other than heating installations. It includes, for example, energy used in water heating, lighting, cooking, refrigeration, clothes washing and drying, dishwashing, and in operating all other appliances. It appears that for most countries, the most important items are water heaters, accounting typically for about 10 percent of total household use, and cooking and refrigeration together, about 7 percent. Lighting of houses (street lighting is included in commercial uses) accounts for only about 1 to 2 percent of total energy consumption, though naturally for a much higher percentage of electricity consumption.

The North American countries consume the most energy for household non-space conditioning uses relative to GDP. The European countries consume an average of about 40 percent less. A new feature is the relatively high consumption level of Japan compared with European countries. Table 4-8, for example, gives data on ownership of selected appliances per 100 households, not the full range of the American household, which would include freezers, clothes driers, and dishwashers, but nonetheless a representative selection. In this table, Japan is shown to be somewhat ahead of European countries in ownership of these appliances, at levels—relative to GDP—comparable to the United States.

There is, no doubt, some difference in the efficiency of appliances among countries, but the major reason for the marked difference between Western European and U.S. levels of energy consumption probably stems from greater use of hot water and the operation of a whole range of appliances—freezers, dishwashers—not commonly used in Europe. By way of illustration: if the United States cut consumption of hot water by half and used only those appliances most commonly used in Europe, that is, those for cooking, lighting, and refrigeration, U.S. consumption of energy for nonheating and cooling uses relative to GDP would fall to Western European levels.

In one sense, because of the relatively small quantities involved, the non-space conditioning category is of little importance in accounting for variations in overall energy/GDP ratios. But in another sense it is of greater importance than the consumption data would suggest, as it is one of the main electricity-consuming sectors.

Relative Income and Price

So far we have partitioned differences in household energy consumption relative to GDP among our countries into differences due to climate, housing characteristics, heating habits, and appliance use. These differences are in turn closely associated with variations in income and in prices of household fuel and energy in our countries. Even variations in

TABLE 4-8 *Energy Consumption for Household Non-space Conditioning Uses, 1972*

	U.S	Canada	France	W. Germany	Italy	Netherlands	U.K.	Sweden	Japan
Consumption of non-space conditioning energy (million tons oil equiv.)	62.6	5.2	5.4	9.0	3.8	1.7	8.5	1.1	13.7
Consumption of non-space conditioning energy (tons oil equiv. per $ million GDP)	53	50	25	37	27	35	45	27	38
Difference from U.S. (tons oil equiv. per $ million GDP)	—	3	28	16	26	18	8	26	15
Stock of selected items of energy-using equipment									
Washing machines (per 100 households)	89	n.a.	65	79	73	85	67	41	98
Refrigerators (per 100 households)	99	n.a.	85	90	91	90	71	92	95
Vacuum cleaners (per 100 households)	97	n.a.	52	86	29	103	85	89	85
Black-and-white TV (per 1,000 inhabitants)	387	n.a.	228	316	192	244	292	356	222
Color TV (per 100 inhabitants)	165	n.a.	19	n.a.	n.a.	48	54	62	139

Source: Data in appendix B.
Note: Blanks = not applicable; n.a. = not available.

FIGURE 4-1   Household Energy Consumption Related to GDP Per Capita, 1972

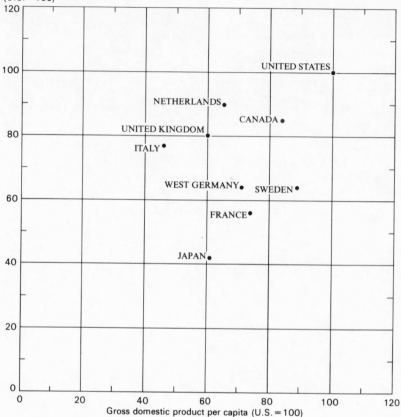

Household energy consumption relative to GDP, corrected for climate (U.S. = 100)

Gross domestic product per capita (U.S. = 100)

climatic conditions cannot be assumed to "explain" variations in heating fuel consumption unless it is assumed that people can afford to buy the amounts of fuel indicated by the severity of the climate.

To take relative income first: it might be thought that because the United States and Canada, the countries with the highest incomes per head, are always among the largest consumers of household energy relative to GDP, there might be some tendency toward a higher consumption of residential fuels and energy relative to income, as income rises. This possibility is explored in figure 4-1, which plots residential energy use (corrected for climatic differences and relative to GDP) against GDP per head. The results are at best ambiguous. The presence of the

United States at the top right-hand corner of the chart suggests some slight tendency for the amount of energy consumed relative to GDP in household uses to rise as per capita income rises. Except for the United States, however, it is difficult to perceive any well-defined tendency for household energy consumption relative to GDP to rise with higher levels of income per head.

The influence of relative prices appears to be more decisive. Table 4-9 gives prices of the main household fuels—coal, gas, heating oil, kerosine, electricity—in all nine countries. With few exceptions, the largest consumers of household energy relative to GDP—United States and Canada—have the lowest prices for all the main household fuels, and Japan—by far the lowest consumer of energy in the residential sector relative to GDP—in almost every case has the highest.

But data on prices of individual fuels do not necessarily give a good indication of the range of energy prices paid *on average* by the household-commercial sectors of the various countries. This is so because the fuel mix varies greatly from country to country.[8] The United States, for example, consumes a relatively large proportion of natural gas, a cheap fuel, but also a relatively large proportion of electricity, a very expensive fuel in terms of thermal content. To arrive at an average "price" per thermal unit for all fuel and power that would take these different fuel mixes into account, two price indexes were calculated (see table 4-9). The pure price index was based on the geometric average of the consumption patterns of the United States and the other countries and, as such, indicates the price of fuel and power based on some idea of an average consumption pattern.

In contrast, the unit value index is weighted only by the consumption pattern of the country concerned and therefore gives an average unit cost of a given thermal amount of household energy in any one country. The difference between the two is instructive. The pure price index shows that the United States is a relatively low-cost country, with prices in many other countries almost twice as high. The unit cost index, however, shows the United States to be only one of a *group* of low-cost countries (including Canada, Netherlands, and Sweden), much closer to the other countries. The difference between energy prices paid in the household category in the United States and Europe may therefore be less than is indicated by the range in prices or the individual fuels.

---

[8] Note that data on fuel mix apply to the total household-commercial sector, whereas price data apply only to the household category of this sector.

TABLE 4-9 *Energy Prices Paid by Households (including taxes), 1972*

Energy type	U.S.	Canada	France	W. Germany	Italy	Netherlands	U.K.	Sweden	Japan
Household category ($ per million kcal)									
Electricity	26.6	32.3	59.7	49.9	51.9	35.3	35.0	25.2	66.6
Gas	4.7	3.8	13.8	12.9	8.0	7.1	10.0	neg.	neg.
Coal	neg.	neg.	11.2	10.4	neg.	neg.	10.7	neg.	neg.
Heating oil	3.1	4.0	5.1	3.7	4.3	4.3	5.9	5.2	4.8
Kerosine	neg.	neg.	neg.	neg.	neg.	8.5	neg.	neg.	9.7
Household category (U.S. = 100)									
Pure price index [a]	100	114.6	205.3	180.9	180.2	150.1	165.4	146.9	190.4
Unit value index [b]	100	109.5	141.1	131.2	132.0	104.2	175.4	110.0	170.0

Note: Household prices deflated by consumption purchasing-power-parity rates of exchange; industry, by GDP purchasing-power-parity rates. Neg. = negligible quantities consumed.

[a] Based on the geometric average of the consumption patterns of United States and other countries.

[b] Weighted by the consumption pattern of the individual country; equals the average unit cost of a given thermal amount of household energy in the individual country.

FIGURE 4-2   *Household Energy Consumption Related to Energy Prices, 1972*

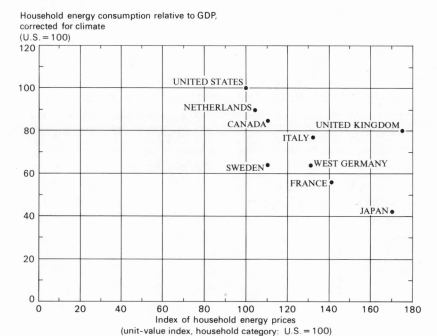

Household energy consumption relative to GDP, corrected for climate (U.S. = 100)

Index of household energy prices
(unit-value index, household category: U.S. = 100)

The relationship between fuel and power prices and energy consumption in the household category relative to GDP is illustrated in figure 4-2, where the unit cost index of fuels and power is plotted against household energy consumption relative to GDP. This plot shows a distinct tendency for countries with high household energy consumption relative to GDP to be associated with low prices, and vice versa. The low-household-energy-price countries—United States, Netherlands, Canada, Sweden—have the larger consumption relative to GDP in this sector, and Japan, with the highest prices, the least. The position of the United Kingdom is somewhat anomalous, having a larger energy consumption relative to GDP than would be suggested by the high energy prices. The reason may be the relatively large amount of coal used for space heating compared with other countries. If consumption data had been expressed in "useful" rather than "total" energy consumed relative to GDP terms, the relative position of the other countries would not have changed much, but the consumption level of the United Kingdom would have been sharply reduced to fall more closely on the trend line (assuming that there was no accompanying change in price).

These observations on the relationship between prices and energy consumed in the household category carry more weight for space conditioning than non-space conditioning. Energy represents only a small part of the total cost of owning and operating the stock of appliances that largely determine this consumption. According to the ICP study, the prices of refrigerators and other household appliances are also relatively cheaper in the United States than in the other countries covered by the ICP study, and the quantities purchased in the United States are substantially higher.[9] Lower relative prices of household energy-using equipment as well as of fuel and power can therefore be assumed to be of significance in explaining the higher U.S. household energy consumption relative to GDP.

A further question is how much of the total variation in energy consumption in the household category relative to GDP can be explained together and separately by variations in price and incomes. Within a limited number of countries, it is not possible to disentangle the relative importance of these different variables, especially as they frequently tend to act in the same direction. Within these nine countries, for example, those which consume the most household energy relative to GDP—United States, Canada, Netherlands, and Sweden—are those with the lowest prices. Three of these countries—United States, Canada, and Sweden—also have the highest per capita income levels, and two have exceptionally rigorous climates. All these variables would be expected to act in the same direction—to increase heating energy consumption. The existence of low prices in *all* of the high consuming countries—including the United States with a low number of degree days and the Netherlands with relatively low income—gives some reason for believing that prices have a key role. Other studies covering more countries or including historical experience have been able, with their more extensive data, to examine more systematically the influence of relative prices and incomes on differences in domestic sector consumption.[10] Because of differences in countries covered, in specifications of the models, and in

---

[9] See chapter 8 of this book for a fuller discussion of this subject.

[10] F. Gerard Adams and James M. Griffin, *Energy and Fuel Substitution Elasticities: Results from an International Cross Section Study,* prepared for United Nations Conferences on Trade and Development (UNCTAD) (New York, UN Economics Research Unit, October 1974); Irving B. Kravis, Zolton Kenessey, Alan Heston, and Robert Summers, *A System of International Comparisons of Gross Product and Purchasing Power* (Baltimore, published for the World Bank by The Johns Hopkins University Press, 1975) chapter 15; William D. Nordhaus, "The Demand for Energy: An International Perspective," in William D. Nordhaus, ed., *Proceedings of the Workshop on Energy Demand* (Laxenburg, Austria, International Institute for Applied Systems Analysis, May 1975).

definitions of income and energy consumption, it is not possible to compare results of these models directly with each other, and with our results.

But a fairly general result emerging from these studies is that when comparing countries, as income rises, consumption of energy in the household sector rises proportionately. With regard to the effect of relative prices, the other studies find the expected relationship between prices and energy consumption; that is, higher relative prices are associated with lower household energy consumption. All of the values of the price elasticities are under 1, and they vary from 0.5 to 0.8.

The prices referred to here include taxes, which vary considerably from country to country. In 1972, for example, taxes on domestic heating fuel in Europe ranged from 10 percent of retail prices in the United Kingdom to 23 percent in Italy, compared with negligible levels in the United States. These differentials in tax levels, reflected in differences in retail prices, suggest that fiscal measures can have a significant influence on levels of energy consumption.

Commercial Category

Table 4-10 gives data on the approximate consumption of energy in the commercial category of the household-commercial sector. The category also includes miscellaneous items. These data must be interpreted with caution as they constitute essentially a residual item derived by deducting estimates of agricultural use, and household use from the OECD "other" sector, which itself is already a residual. This "other" sector is known to include a wide variety of frequently unrelated energy-using areas—schools, offices, restaurants, apartment buildings, artisans' shops, department stores, small-scale construction sites, government buildings, street lighting, and also some military uses—but it is by no means certain that coverage is consistent between countries. In some countries, for example, much of military use may be included in stock changes in industrial use, or in transport; while in others, all may be included under the "commercial including miscellaneous" category (hereafter in our study termed "commercial").

With these reservations, then, table 4-10 shows this category to account for approximately 10 percent of total energy consumption in all countries. The United States and Canada consume 30 percent more energy in commercial use relative to GDP than the European countries and about twice the Japanese level. This category accounts for approximately 7 percent of the total difference in energy/GDP ratios among countries.

TABLE 4-10 *Energy Consumption in the Commercial Category, 1972*

Energy consumption	U.S.	Canada	France	W. Germany	Italy	Netherlands	U.K.	Sweden	Japan
Commercial category (million tons oil equiv.)	155.0	14.7	18.1	31.7	10.3	5.9	15.2	4.3	26.1
Commercial category relative to GDP (tons oil equiv. per $ million GDP)	131	142	84	129	72	120	80	106	72
Difference from U.S. (tons oil equiv. per $ million GDP)	—	−11	47	2	59	11	51	25	59
Difference from U.S. after adjustment for climate [a] (tons oil equiv. per $ million GDP)	—	14	43	9	45	20	43	47	51

Note: The commercial category includes "miscellaneous." Blanks = not applicable.

[a] Climatic adjustment made by deducting 10 mtoe of the total 1972 U.S. energy consumption of 13.5 mtoe in commercial air conditioning to represent the hotter U.S. climate. The remainder was then adjusted for different winter climates by standardizing on the basis of one-half of the difference in degree days between the United States and the other country. One-half rather than the full total was used to take into account the diminished influence of ambient temperature on the heating needs of commercial buildings.

Despite the wide variety of activities included in this sector, most of them share an important characteristic—about 80 percent of the energy consumed is for conditioning the space in buildings. As such, differences in climate might be expected, as in the household category, to play some part in explaining differences in energy consumption in commercial uses relative to GDP between countries. In some respects, however, the heating and cooling patterns of the commercial category differ from those of the household category. A large part of the effective heating in offices comes from lighting and other nonheating functions. Air conditioning is used in all countries (even where it is not strictly necessary in terms of outside climate) as part of the managed climatic environment inherent in newer office building design throughout the world. Thus, the space conditioning needs of office buildings are not as closely related to ambient temperature as is the case with household heating consumption. To take this into account, only one-half of the total degree day adjustment has been used. As in the case of the household category, the United States is, after climatic adjustment, a higher consumer than the other countries.

It is difficult to assess what part of this total adjusted difference in U.S. consumption in the commercial category is caused by the size of the sector and what part is caused by the energy intensity of commercial activities. There is no satisfactory measure of overall output for the commercial category, and even such indicators as square footage of office space were not available for all our countries. Schipper and Lichtenberg in their study of Swedish and U.S. energy consumption show commercial energy consumed per square meter of office space to be some 30 percent lower in Sweden even before the difference in heating degree days is taken into account.[11] If this result were generalized to the other European countries, it might be implied that most of the higher U.S. energy consumption in commercial use relative to GDP is due to higher intensity of energy use rather than relative sector size or type of activity. But for other countries, the larger size of the U.S. commercial category would account for some of the difference.

Some indication of the size of the commercial category is indicated by the value of GDP originating in some important branches of activity —wholesale and retail trade; restaurants; hotels; storage; communication; insurance; real estate; business services; and community, social,

[11] Schipper and Lichtenberg, *Efficient Energy*.

and personal services.[12] According to these data, the United States has one of the larger service sectors, accounting for about 45 percent of the total, compared with about one-third for most countries. This would imply that, in general, the large size of the U.S. commercial category accounts for much of the higher U.S. consumption of energy relative to GDP in the commercial sector.

Agricultural Category

As already noted, the direct consumption of energy in the agricultural category is a relatively small proportion (about 5 to 10 percent) of the household-commercial sector consumption, and therefore an even smaller proportion of total energy consumption. Only two countries—the Netherlands and Canada—consume as much as 3 to 5 percent of total energy in agriculture; many of the others consume less than 1 percent.[13]

Despite this surface similarity, marked differences in energy consumption for agriculture relative to GDP do exist (see table 4-11). For example, the United States, Canada, and the Netherlands consume more than twice as much energy for agricultural purposes relative to GDP as other countries. In the case of the Netherlands, the large consumption is undoubtedly strongly affected by the output mix. This country has an exceptionally large horticultural sector requiring large amounts of energy —particularly natural gas—for heating greenhouses. For the other two countries, no such clear-cut reason emerges; but, in general, it seems that neither sector size nor the little we know about the mix effect goes far in explaining the high U.S. (and Canadian) energy consumption in agriculture relative to GDP. Rather the reverse: if anything, these effects tend to widen the gap between the United States and other countries, suggesting that the United States and Canada are very much more energy intensive in their agricultural practices. But an important offsetting consideration needs to be borne in mind.

Our energy consumption data must be used with reserve in making intercountry comparisons, because agriculture is more prone than most

---

[12] Note that the standard industrial code (SIC) industrial classification upon which this is based hides many contrasts in the way commercial jobs are recorded; to the extent that, say, the Japanese steel industry internalizes its accountancy, advertising, wholesale activities, transport fleets, and community services, the apparent size of their commercial sector is smaller.

[13] OECD energy consumption data refer only to direct-energy uses and exclude energy embodied in fertilizer and other agricultural inputs, which in our statistical format would be included under energy consumption in industrial uses.

TABLE 4-11 *Energy Consumption in the Agricultural Category and Related Data, 1972*

Energy consumption	U.S.	Canada	France	W. Germany	Italy	Netherlands	U.K.	Sweden	Japan
Energy consumption in agriculture (million tons oil equiv.)	29.0	5.0	2.8	1.9	2.0	2.1	2.0	0.5	2.7
Energy consumption in agriculture (tons oil equiv. per $ million GDP)	25	48	13	8	14	43	11	12	7
Gross domestic product originating in agriculture,[a] as percentage of total GDP	3.0	3.6	6.0	3.0	8.4	(5.0)	2.5	3.7	5.4
Agricultural labor force[b] (thousands)									
Including unpaid family workers	3,472	574	2,635	1,953	3,298	316	n.a.	287	7,050
Excluding unpaid family workers	3,001	463	n.a.	908	2,591	n.a.	739	232	3,500
Energy consumption per worker (tons oil equiv.)									
Including unpaid family workers	8.35	8.71	1.06	0.97	0.61	6.65	n.a.	1.74	0.38
Excluding unpaid family workers	9.66	10.80	n.a.	2.09	0.77	n.a.	2.71	2.16	0.77

Note: Figure in parentheses is RFF estimate; n.a. = not available.
Sources: Data for gross domestic product originating in agriculture are from Organisation for Economic Co-operation and Development, *National Accounts of OECD Countries, 1961–72.* Data for agricultural labor force are from Organisation for Economic Co-operation and Development, *Labor Force Statistics, 1961–72.*
a Includes hunting, forestry, and fishing.
b Civilian employment.

sectors to problems of definition and comparability. The nature of agricultural enterprise varies considerably from country to country. In Italy and Japan, for example, unpaid family workers account for 21 to 50 percent of the total agricultural work force compared with under 1 percent for Canada and the United States. There is also some presumption that energy consumption data for those countries with a more formally organized agricultural sector will be more comprehensive than data relating to other forms of organizations where farming may be combined with other activities. In this way, the agricultural energy consumption of some of the European countries and Japan may be underestimated compared with that of the United States and Canada.

Even so, it seems likely that the U.S. agricultural sector uses substantially more direct energy to produce a similar output than does Western Europe. This higher energy intensity almost certainly reflects a greater degree of mechanization and productivity per person than in Western Europe. In any event, the small amount of energy consumed in the agricultural category—only half as much again as in air conditioning—accounts for only a small part of the total variation in energy consumption between countries.

Summary

In this chapter we have analyzed energy consumption relative to GDP in the household-commercial sector. For many of our countries, this sector accounts for between 20 and 30 percent of the total difference in energy/GDP ratios, compared with the United States. Unlike the transport sector, where energy consumption is very similar in all European countries, there is considerable variation within Europe in the household-commercial sector, with the Netherlands and Sweden having comparatively high energy consumption relative to GDP. Consequently, it is more difficult to generalize in this sector between North American and Western European patterns.

Within the household-commercial sector, the main difference lies in household space conditioning. After correction for climatic differences, the United States emerges as the largest consumer of heating and cooling energy relative to GDP, with a consumption level generally 40 percent higher than many Western European countries. It is estimated that about one-half of this difference can be attributed to the prevalence of the larger, single family house. The other half is attributable to different heating habits, such as the heating of unoccupied rooms, starting to heat at relatively high temperatures, and the like. The effect of higher insulation standards is assumed to be significant only when the United States is compared with Sweden. The efficiency of heating plants, higher in the United States, is not responsible for any of the differences between countries.

The quantification of the effect of these factors may be misleading. Thus for the United States, the adjustment for climate may contain an element of different heating and cooling habits, and the adjustment for size of house will contain some elements of superfluous heating. There is, in short, considerable overlap among these variables that it was not possible to allow for in the results.

Nor was it possible, owing to lack of adequate output data, to partition the energy consumption of the other parts of this sector—non-space conditioning, agricultural, and commercial—in an analogous manner. The greater U.S. consumption in commercial uses relative to GDP is believed to be due in many cases to a larger service sector in that country. The mix and intensiveness effects will also play some part. The variation in agricultural consumption relative to GDP—very marked between countries but with little significance for the overall energy GDP ratio—appears to be largely due to a greater direct-energy input per unit of output.

The existence of such variation in energy consumption between countries of similar economic structure raises the question of whether, in the household-commercial sector, particularly, there might not be opportunities for substantial energy savings in the United States without radical changes in life-style or substantial expenditures. Or, put in a different way, would substantial energy savings entail radical changes in life-styles?

To illustrate a range of possibilities we consider three cases:

1. The first case would be one in which energy saving could be achieved with minimum disruption to personal life-style and with minimum expenditure. In terms of our analysis, this would be a change in "excessive" heating habits. It would involve such measures as starting heating at lower temperatures, reducing the heating of unoccupied spaces, and lowering the thermostat.

According to our analysis, such actions could reduce household consumption for heating purposes by about 18 percent and commercial consumption for heating purposes by about 10 percent. Similarly, a reduction in hot water use by one-half and a less intensive use of domestic appliances—perhaps a more painful life-style adjustment than turning down a thermostat—would reduce U.S. energy consumption in non-space conditioning uses by about 50 percent. Together, these changes could, theoretically, result in a 15 percent reduction in household-commercial energy use, though in practice actual savings might well be smaller.

2. A second case would distinguish the savings that could be made without any change in living patterns but at some initial cost and over a longer period of time by the use of better insulation. Our analysis, covering mainly countries which by and large do not have better insulation than the United States, is not well suited to providing a systematic answer to this question.[14] However, the inclusion of Canada and Sweden, whose insulation standards are superior to those of the United States, does give some indication of how much those savings could be. It will be recalled that Swedish U-values are typically one-half the value of those of the United States and other European countries. This implies that heat loss, and therefore energy consumption for households, could be reduced substantially through improved insulation. Controlled studies done in this country on the effects of improved insulation in new houses confirm the possibility of such savings. A 50 percent savings in space conditioning fuel consumption would represent a 40 percent reduction in household and commercial energy consumption. Savings of this magnitude are unlikely to be achieved in practice, particularly in the shorter term, as retrofitting of existing houses saves less energy than comprehensive insulation of new housing. A recent OECD report, for example, estimates the savings to be achieved in the United States and several European countries by improved insulation in the household-commercial sector at an average of nearer 20 percent.[15]

3. A final case could be the energy savings to be achieved from substantial changes in life-styles. Following our analysis, this might represent for the household sector, the move from single-family houses into multifamily dwellings. Our study indicates that this might save approximately 18 percent of household consumption of heating fuels. If a move to multifamily houses were to be accompanied by a move away from the suburbs into more compact urban communities, the greater energy sav-

[14] Similarly, it has not been possible to take into account the effect of different heating systems—district heating, solar energy, heat pumps—which are not yet widely used in any of the countries, with the exception of district heating in Sweden, but which promise much higher efficiency than existing systems.

[15] Organisation for Economic Co-operation and Development, *OECD World Energy Outlook* (Paris, OECD, 1977).

ings would, of course, be in transport rather than in heating fuel consumption.

Even without such radical life-style changes, an improvement in insulation plus a change in heating habits could make a substantial reduction in U.S. household-commercial consumption. If only one-third of the possible insulation improvements and one-half of the heating habit changes were achieved, energy consumption in the household-commercial sector could fall by more than 15 percent. A less intensive use of appliances and hot water could push up this saving to nearer 20 percent.

CHAPTER 5

Transport Sector

THIS CHAPTER EXPLORES the extent to which the use of energy in the transport sector of the economy contributes to the differences in overall energy/GDP ratios between the United States and the other countries included in our study. In the first part of the chapter there is a brief, comparative overview of transportation energy use as a whole. The second—and major—part deals with energy consumption for passenger transport. Compelled by the nature of the study to deal selectively, rather than exhaustively, with the various components of energy consumption, we deemed it important to accord particular attention to passenger transport. Perhaps more than any other energy-using activity, it symbolizes, in the minds of many, the gap between "efficient" (read foreign) and "inefficient" (read U.S.) consumption patterns. And no wonder: the contrast between, say, large and small cars or between rail and automotive travel often endures as a vivid personal legacy in the recollections of those who have travelled abroad. The third part of the chapter presents some observations on freight transport. A short concluding section ties the different strands together.

Transportation As a Whole

In 1972, energy used in transportation accounted for 22 percent of America's nationwide energy consumption. For Canada, the share was 17 percent; for other countries, decisively lower still. For the six Western European countries in our study, the share averaged 13 percent. In our earlier comparative review of the principal components of energy demand, we pointed to transport as being—in virtually every one of our eight pairs of comparisons with the United States—the single most important contributor to overall energy/GDP variability. The household-commercial sector ranks as the next most important. But even with its many categories, the household-commercial sector tends to account for a substantially lower proportion of the energy/GDP difference than does the transport sector.

A broad quantitative perspective on total transportation energy is shown in table 5-1. To facilitate a viewing of transportation energy within its relevant economic and demographic context, a number of indicators in this table are repeated from earlier tables in the study. The conspicuous role of transport in overall energy/GDP differentials emerges clearly from the table. The minimum contribution that the transport sector makes to energy/GDP differences with the United States is as much as 30 percent in the United States–France comparison; the maximum is over 90 percent in the United States–Netherlands comparison. For the group of six Western European countries, the figure is 40 percent.[1]

As one looks at the per capita indexes across the bottom of the table, the same phenomenon shows up in a slightly different way. Relative to the United States, GDP per capita is typically the highest of the indexes, followed by energy per capita, and transport energy per capita in that (sharply falling) order. Although there are exceptions, per capita consumption of energy for passenger transport in other countries tends, for the most part, to be markedly below per capita levels of the freight category, compared to their respective U.S. levels.[2]

As a result, the indexes measuring energy/GDP ratios relative to the United States are distinctly lower for passenger transport energy use than for total energy use. For the six Western European countries, for example, the overall energy/GDP ratio is two-thirds of that for the United States; the passenger transport energy/GDP ratio is less than two-fifths the U.S. level. Two extremes surrounding the Western European case are Canada, which discloses a picture reasonably close to that of the United States; and Japan, where the spread (relative to the respective U.S. ratios) between the overall energy/GDP ratio and the passenger transport energy/GDP ratio is widest of all. Clearly, passenger transportation deserves close scrutiny in any effort to illuminate comparative international energy-use patterns.

---

[1] These percentages should, however, be understood to coexist in some cases with offsetting negative numbers. In the case of Netherlands, for example, the positive percentage distribution of differences from the U.S. energy/GDP ratio adds up to 177 percent, offset by −16 percent for households and −61 percent for nonenergy uses (see table 3-3). Also, the six-nation Western European grouping is, of course, something of an abstraction: the characteristics of the Italian economy, after all, do not correspond to Sweden's. However, the combined treatment does have the expositional virtue of capturing in a succinct fashion some of those essential features that are broadly representative of numerous foreign countries. Moreover, serious data deficiencies and uncertainties for given countries may wash out in the aggregated computation.

[2] The freight category also includes school bus transport, flights in personal or company aircraft, and personal use of trucks. See footnote a to table 5-1.

TABLE 5-1   Energy Consumption in the Transport Sector and Related

Item	U.S.	Canada	France	W. Germany
GDP (billion dollars)	1,178.49	103.31	215.47	246.10
Population (millions)	208.84	21.85	51.70	61.67
Total energy consumption				
(million tons oil equiv.)	1,744.65	183.09	171.23	253.85
Transport	385.11	31.55	25.30	32.59
Passenger	256.28	17.92	15.12	20.73
Freight ª	128.83	13.63	10.18	11.86
GDP per capita (thousand dollars)	5.643	4.728	4.168	3.991
Total energy consumption per				
capita (tons oil equiv.)	8.354	8.379	3.312	4.116
Transport	1.844	1.444	0.489	0.528
Passenger	1.227	0.820	0.292	0.336
Freight ª	0.617	0.624	0.197	0.192
Transport as percent of total				
energy consumption	22.1	17.2	14.8	12.8
Passenger	14.7	9.8	8.8	8.2
Freight ª	7.4	7.4	5.9	4.7
Total energy/GDP ratio (tons				
oil equiv. per $ million GDP)	1,480	1,772	795	1,031
Transport	327	305	117	132
Passenger	218	173	70	84
Freight ª	109	132	47	48
Difference between U.S. and				
other countries' energy/				
GDP ratios (tons oil equiv.				
per $ million GDP)	—	−292	685	449
Transport	—	22	210	195
Passenger	—	44	147	133
Freight ª	—	−23	62	61
Percent distribution of energy/GDP				
differences:				
Total energy consumption	—	ᵇ	100.0	100.0
Transport	—	ᵇ	30.7	43.4
Passenger	—	ᵇ	21.5	29.6
Freight ª	—	ᵇ	9.1	13.6
GDP per capita				
(index, U.S. = 100)	100.0	83.8	73.9	70.7
Total energy consumption per capita				
(index, U.S. = 100)	100.0	100.5	39.6	49.3
Transport	100.0	78.3	26.5	28.6
Passenger	100.0	66.8	23.8	27.4
Freight ª	100.0	101.1	31.9	31.1
Total energy/GDP ratio				
(index, U.S. = 100)	100.0	119.7	53.7	69.7
Transport	100.0	93.3	35.8	40.4
Passenger	100.0	79.7	32.3	38.7
Freight ª	100.0	121.1	43.1	44.0

Note: Blanks = not applicable.

Sources: GDP, population, total energy consumption, and transport energy are from data in appendix B.
The freight transport figure is a residual. The derivation of passenger transport is discussed in appendix D.

Data, 1972

Italy	Netherlands	U.K.	Sweden	Six W. European countries	Japan
142.13	49.03	190.03	40.65	883.41	362.71
54.41	13.33	55.88	8.13	245.12	105.97
130.01	62.36	213.10	43.19	873.74	307.82
19.36	6.57	27.49	4.91	116.22	38.07
15.42	3.73	15.63	3.57	74.20	17.38
3.94	2.84	11.86	1.34	42.02	20.69
2.612	3.678	3.401	5.000	3.604	3.423
2.389	4.678	3.814	5.312	3.565	2.905
0.356	0.493	0.492	0.604	0.474	0.359
0.283	0.280	0.280	0.439	0.303	0.164
0.072	0.213	0.212	0.165	0.171	0.195
14.9	10.5	12.9	11.4	13.3	12.4
11.9	6.0	7.3	8.3	8.5	5.6
3.0	4.6	5.6	3.1	4.8	6.7
915	1,272	1,121	1,062	989	849
136	134	145	121	132	105
108	76	82	88	84	48
28	58	62	33	48	57
565	208	359	418	491	631
191	193	182	206	195	222
109	141	135	129	134	169
81	51	47	76	61	52
100.0	100.0	100.0	100.0	100.0	100.0
34.3	92.8	50.7	49.3	39.7	35.2
19.6	67.8	37.6	30.9	27.2	26.8
14.6	24.5	13.1	18.2	12.4	8.2
46.3	65.2	60.3	88.6	63.9	60.6
28.6	56.0	45.7	63.6	42.7	34.8
19.3	26.7	26.7	32.8	25.7	19.5
23.1	22.8	22.8	35.8	24.7	13.4
11.7	34.5	34.4	26.7	27.7	31.6
61.8	85.9	75.7	71.8	66.8	57.4
41.6	41.0	44.3	37.0	40.4	32.1
49.8	35.0	37.8	40.6	38.5	22.1
25.7	53.2	56.9	30.3	44.0	52.3

[a] Properly termed "freight and other," this category also includes school bus transport, flights in personal or company aircraft, and personal use of trucks. In tables 5-8 and 5-9, a narrower and more precise definition of freight is employed.

[b] Calculation not meaningful.

Passenger Transport

Ascribing to the transport sector, and more specifically to passenger transport, an important reason for variability in energy/GDP ratios does not represent any deep analytical probing. In fact, it reflects no more than the process of dividing each country's energy-use sectors (for example, transport) by a given number—in this case, by each country's GDP. The exercise thus serves only to identify, and to alert us to, those aspects of energy consumption which we need to penetrate much more searchingly if we are to offer useful insight into the nature of international energy/GDP differences.[3] In other words: what are the factors at work that cause the energy used in passenger transport, relative to GDP, to be strikingly lower in Western Europe and Japan than in North America? In what follows, we will first set out to dissect the problem statistically; we will then touch on possible explanations for the differential passenger transport characteristics revealed by the data.

Estimates of two sets of statistical indicators are necessary for this analysis: (1) each country's aggregate number of passenger-miles, distributed by the major travel modes (cars, buses, rail, and air); and (2) each country's aggregate consumption of energy for purposes of passenger transport—again, split among the respective travel modes. (The latter divided by the former give energy intensities in the aggregate or for particular modes.) Any insight into comparative patterns of travel and energy use depends, at a minimum, on the availability and tolerable quality of such data and on how they are linked to other appropriate economic and demographic indicators. (Derivation of the passenger mileage and fuel consumption figures, and their reliability, are matters of considerable difficulty. Appendix E deals with the problem at some length.)

Table 5-2 summarizes comparative passenger-mile and energy-use figures, while table 5-3 presents underlying detail by passenger mode. At the aggregative level portrayed in table 5-2, we observe that the per capita travel activity of Americans greatly exceeds that of other countries. On the average, per capita passenger mileage among Americans is three times the Japanese figure and over twice that of most European countries. Per capita travel in Canada is—not surprisingly given the conti-

---

[3] Indeed, we should also be interested in those ratios which are less than statistically significant because such ratios can obscure genuine differences in the intensity or economic efficiency with which energy is used, offset by the intrasectoral "product mix" differences among countries. Specific cases that illustrate this point will emerge as the chapter progresses.

TABLE 5-2 Summary of Passenger Transport Data, 1972

Countries	Passenger travel			Passenger transport energy consumption				
	Million passenger-miles (1)	Passenger-miles per capita (2)	Passenger-miles per thousand $ GDP (3)	Million tons oil equiv. (4)	Tons oil equiv. per capita (5)	Tons oil equiv. per $ million GDP (6)	Tons oil equiv. per hundred thousand passenger-miles (7)	Dollars of GDP per capita (8)
U.S.	2,357,349	11,288	2,000	256.28	1.227	218	10.872	5,643
Canada	143,200	6,554	1,386	17.92	0.820	173	12.515	4,728
France	205,500	3,975	954	15.12	0.292	70	7.358	4,168
W. Germany	362,270	5,874	1,472	20.73	0.336	84	5.722	3,991
Italy	226,500	4,163	1,594	15.42	0.283	108	6.808	2,612
Netherlands	61,592	4,621	1,256	3.73	0.280	76	6.060	3,678
U.K.	278,800	4,989	1,467	15.63	0.280	82	5.605	3,401
Sweden	51,073	6,282	1,256	3.57	0.439	88	6.988	5,000
Six Western European countries	1,185,735	4,837	1,342	74.20	0.303	84	6.258	3,604
Japan	398,284	3,758	1,098	17.38	0.164	48	4.365	3,423
	Index, U.S. = 100							
U.S.		100.0	100.0		100.0	100.0	100.0	100.0
Canada		58.1	69.3		66.8	79.7	115.1	83.8
France		35.2	47.7		23.8	32.3	67.7	73.9
W. Germany		52.0	73.6		27.4	38.7	52.6	70.7
Italy		36.9	79.7		23.1	49.8	62.6	46.3
Netherlands		40.9	62.8		22.8	35.0	55.7	65.2
U.K.		44.2	73.4		22.8	37.8	51.6	60.3
Sweden		55.7	62.8		35.8	40.6	64.3	88.6
Six Western European countries		42.9	67.1		24.7	38.5	57.6	63.9
Japan		33.3	54.9		13.4	22.1	40.1	60.6

Sources: The derivation of passenger-mile and energy consumption data is described in appendix D. GDP and population data are from appendix B.

nental breadth of the country involved—closest to that in the United States, but even so reflects an index that is 42 percent lower.

Since the dispersion of per capita GDPs spans a much narrower range than per capita travel, the ratio of passenger-miles relative to GDP between the United States and other countries varies markedly. From just this cursory look at the data, we can see that the passenger transportation sector's contributions to the high U.S. energy/GDP ratio may at least in part be the consequence of Americans' propensity to travel— irrespective of how much energy it takes.

In particular, America's high ratio of passenger-miles to GDP reflects a correspondingly high U.S. ratio of car ownership to GDP. Per $100,000 of GDP, there were, in 1972, 8.2 automobiles registered in the United States, compared with 7.2 for Canada, a 6.0 to 6.5 range for most of Western Europe, and 3.4 for Japan. A surprising exception is Italy, which shows as high a ratio of cars on the road relative to GDP as does the United States. The fact that Italy's passenger-mile/GDP ratio nonetheless falls below the United States must implicitly be due to lower load factors and miles per vehicle, or both.

And, of course, U.S. passenger transport *is*, in the aggregate, highly energy intensive, as column (7) of table 5-2 indicates. The U.S. energy/passenger-mile ratio does fall below the Canadian figure, but it is decisively above that for all the other countries. As a matter of simple arithmetic, therefore, the combination of a travel-intensive GDP and a situation of energy-intensive passenger travel confers on the United States the high ratio of passenger transport energy consumption to GDP which we have already commented on.

Anatomy of Intercountry Differences

We now continue our country comparisons by probing the characteristics of their transport modes. The necessary data appear in table 5-3 where two items are particularly important: the country differences in modal mix, and the energy intensities (that is, energy consumption relative to passenger mileage) characteristics of the respective modes.[4] These

---

[4] There were formidable measurement problems connected with the estimation of these indicators. In particular, the reliability of the energy intensity figure is dependent on the reliability of the numerator (energy consumption) and denominator (passenger-miles), of which it is the quotient. Since numerator and denominator frequently emanated from independent sources, reliable and consistent treatment could not be ensured. Note also that the intensity measure for any given mode implicitly reflects a number of contributory elements each of which has a bearing on the indicated ratio and each of which, in some sense, is connected with the matter of "efficient" usage. Examples include load factors and intraurban versus interurban

two factors, along with country-by-country differences in the volume of travel (relative to GDP), already touched upon, are central to further discussion. The proportion of passenger-miles represented by automotive transport is highest for the United States (92 percent), lowest for Japan (an astonishing 34 percent). Among the countries, the Canadian automotive share is closest to that of the United States, with 88 percent; while for most of the Western European countries, the figure hovers around 80 percent. The addition of domestic air travel, which is minimal within our non–North American countries, skews these intercountry proportions even further.[5] Conversely, the comparative shares of passenger travel accounted for by public ground transportation—bus and rail systems— range from a low of 3 percent in the United States to a high of 64 percent in Japan. The typical Western European figure is close to 20 percent.

As table 5-3 indicates, the modes of travel that attain their highest national share in the United States (cars and air) are characterized by relatively high energy intensities, while forms of public transport that are more important in Europe and Japan have relatively low intensities. Moreover, for at least one transport mode—the critically important automotive component—the energy intensity in the United States is substantially above that overseas.[6]

Having identified three key aspects (passenger-mileage relative to GDP, modal mix, and energy intensity) in the determination of the comparative ratios of energy consumption in passenger transport to aggregate GDP, we now assess their relative importance in a more systematic fash-

---

transport. (Identical passenger cars can obviously yield substantially different intensities if the first is driven at a low load factor on city streets and the second is driven at a high occupancy rate on the highway.) Ideally, one would want to show how the intensity measure we have is affected by such factors. But in the multicountry and statistically problematic circumstances with which we had to grapple, such further partitioning was simply not feasible. These, and numerous other issues, are discussed in appendix D.

[5] If we were able to assign to nationals of different countries their volume of *international* air travel, the non–North American air proportions would no doubt rise. The Dutch fly to Nice but not between Rotterdam and Amsterdam. Americans or Canadians, by contrast, do proportionately more of their flying internally. See comments in appendix D.

[6] Intercountry variations in intensity for given modes should be regarded with a certain amount of circumspection. (For example, the Canadian rail intensity seems inexplicably higher than the U.S. In fact, this is due both to the lower load factor in Canadian intercity rail traffic and to the weight, in the U.S. statistics, of the low-intensity New York City subway system.) Some of the difficulties in measurement and interpretation are discussed in appendix D. In general, however, we believe large differences among countries to be genuine. Averaging by groups of countries helps.

TABLE 5-3 *Passenger-Miles, Energy Consumption and Related Data, by Travel Mode, 1972*

Travel mode	Passenger transport energy consumption		Passenger-miles		Energy consumption (tons oil equivalent per 100 thousand passenger-miles)	Passenger-miles (per $ thousand GDP)	Energy/GDP ratio (tons oil equivalent per $ million GDP)
	Million tons oil equiv.	Percentage	Millions	Percentage			
United States							
Cars	226.92	88.5	2,171,332	92.1	10.450	1,842	192.6
Buses	1.74	0.7	48,955	2.1	3.550	42	1.5
Rail	0.84	0.3	17,759	0.8	4.758	15	0.7
Air	26.78	10.4	119,303	5.1	22.450	101	22.7
Total	256.28	100.0	2,357,349	100.0	10.872	2,000	217.5
Canada							
Cars	16.07	89.7	125,700	87.8	12.782	1,217	155.6
Buses	0.30	1.7	8,800	6.1	3.380	85	2.9
Rail	0.28	1.6	2,500	1.7	11.000	24	2.7
Air	1.28	7.1	6,200	4.3	20.645	60	12.4
Total	17.92	100.0	143,200	100.0	12.515	1,386	173.5
France							
Cars	13.13	86.8	159,120	77.4	8.250	738	60.9
Buses	0.45	3.0	15,525	7.6	2.900	72	2.1
Rail	0.53	3.5	26,355	12.8	2.000	122	2.5
Air	1.01	6.7	4,500	2.2	22.500	21	4.7
Total	15.12	100.0	205,500	100.0	7.358	954	70.2
West Germany							
Cars	17.50	84.4	297,000	82.0	5.892	1,207	71.1
Buses	1.27	6.1	39,000	10.8	3.250	158	5.2
Rail	1.46	7.0	24,600	6.8	5.935	100	5.9
Air	0.50	2.4	1,670	0.5	30.000	7	2.0
Total	20.73	100.0	362,270	100.0	5.722	1,472	84.2
Italy							
Cars	13.68	88.7	180,000	79.5	7.600	1,266	96.2
Buses	0.55	3.6	22,000	9.7	2.482	155	3.9
Rail	0.44	2.9	22,000	9.7	2.020	155	3.1
Air	0.75	4.9	2,500	1.1	30.000	18	5.3
Total	15.42	100.0	226,500	100.0	6.808	1,594	108.5

Netherlands							
Cars	3.44	92.2	50,080	81.3	6.872	1,021	70.2
Buses	0.19	5.1	6,520	10.6	2.842	133	3.9
Rail	0.11	2.9	4,992	8.1	2.125	102	2.2
Air	neg.	neg.	neg.	neg.	neg.	neg.	neg.
Total	3.73	100.0	61,592	100.0	6.060	1,256	76.1
United Kingdom							
Cars	12.75	81.6	222,200	79.7	5.735	1,169	67.1
Buses	1.24	7.9	34,200	12.3	3.632	180	6.5
Rail	1.25	8.0	21,100	7.6	5.922	111	6.6
Air	0.39	2.5	1,300	0.5	30.000	7	2.1
Total	15.63	100.0	278,800	100.0	5.605	1,467	82.3
Sweden							
Cars	3.21	89.9	43,000	84.2	7.470	1,058	79.0
Buses	0.14	3.9	4,000	7.8	3.548	98	3.4
Rail	0.12	3.4	3,673	7.2	3.402	90	3.0
Air	0.09	2.5	400	0.8	22.500	10	2.2
Total	3.57	100.0	51,073	100.0	6.988	1,256	87.8
Six Western European countries							
Cars	36.71	85.9	951,400	80.2	6.696	1,077	72.1
Buses	3.84	5.2	121,245	10.2	3.167	137	4.3
Rail	3.91	5.3	102,720	8.7	3.806	116	4.4
Air	2.74	3.7	10,370	0.9	26.422	12	3.1
Total	74.20	100.0	1,185,735	100.0	6.258	1,342	84.0
Japan							
Cars	11.81	68.0	136,806	34.3	8.630	377	32.6
Buses	1.78	10.2	67,192	16.9	2.648	185	4.9
Rail	2.33	13.4	186,486	46.8	1.250	514	6.4
Air	1.46	8.4	7,800	2.0	18.750	22	4.0
Total	17.38	100.0	398,284	100.0	4.365	1,098	47.9

Note: neg. = negligible.

Sources: The derivation of passenger-mile and energy consumption data is described in appendix D. GDP data are from appendix B.

ion. A detailed discussion of our procedure appears in appendix E; high-lights appear in table 5-4.

What we have attempted to do in table 5-4 is to partition the variability between the United States and other countries in passenger transport energy/GDP ratios (the last column of table 5-3) into three elements: a "modal mix effect," which relates to the effect of varying distributions among the countries of cars, buses, rail, and air transport; an "activity mix effect," which refers to the fact that, relative to GDP, more passenger-miles of transport are generated in one country than another; and "intensity factors," which reflect the amount of energy associated with a passenger-mile of travel by a given mode. (The so-called interaction columns, shown in the table, arise from the fact that in relating these three elements to their joint aggregate effect, we are dealing with multiplicative rather than additive components; see appendix E for a fuller explanation.)

Interpretation of the table is fairly straightforward. The last column of the Netherlands line, for example, tells us that, per million dollars of GDP, the United States uses 141 tons oil equivalent of energy more than does Holland for passenger transport. By far the most important underlying reason for that difference (81 tons oil equivalent) lies in the activity mix effect: relative to income, we travel more than do the Dutch. Next most important (accounting for 67 of the 141 tons oil equivalent difference) is intensity, presumably reflecting principally the smaller size of Dutch cars, but also perhaps attributable to other factors, such as passenger loads carried. Finally, the modal mix factor (the greater propensity toward use of rail and buses in Holland, as depicted in table 5-3) explains only half as much as the intensity component. Another way of posing the question suggested by these figures is: how much would U.S. energy consumption in passenger transportation decline relative to the U.S. GDP if Dutch transport characteristics applied to this country? To the extent that the relatively low volume of travel associated with a small country like Holland is inapplicable to the United States, the question is largely meaningless. To the extent that it forces us to think, say, about altered travel modes or more efficient cars, the question may be less fanciful. More on this point later.

A glance at table 5-4 shows that in five of the binary comparisons (Canada, France, Netherlands, Sweden, and Japan) the activity mix exerts the greatest impact on the difference from the United States. For West Germany, Italy, and United Kingdom, intensity factors dominate. Proportionately, the modal mix effect is most important in the case of

TABLE 5-4 *Analytical Breakdown of Differences Between United States and Other Countries in Passenger Transport Energy/GDP Ratios, 1972*

(tons oil equivalent per $ million GDP)

| Countries | Due to mix factors | | | | Due to intensity factors | Due to interaction between total mix and intensity | Total actual difference (U.S. minus country shown) |
	Modal mix effect	Activity mix effect	Interaction	Total mix effect			
Canada	8.7	109.0	−4.5	113.2	−42.0	−27.2	44.0
France	28.2	113.7	−14.7	127.2	41.2	−21.1	147.3
W. Germany	29.7	57.5	−7.9	79.3	76.3	−22.3	133.3
Italy	30.3	44.4	−6.4	68.3	45.8	−5.1	109.0
Netherlands	32.3	80.9	−12.0	101.2	66.7	−26.5	141.4
Sweden	25.5	80.8	−9.4	96.9	55.1	−22.3	129.7
U.K.	32.7	58.0	−8.6	82.1	79.1	−26.0	135.2
Six Western European countries	30.3	71.6	−10.0	91.9	65.5	−23.9	133.5
Japan	80.3	98.1	−36.3	142.1	38.2	−10.7	169.6

Source: Data in table 5-3. The methodology is discussed in appendix E.

the United States–Japan comparison: it arises from the ranking importance of Japanese railways, but even for Japan, the modal mix effect trails
the activity mix effect. Note that in all twenty-four comparisons in table
5-4 (three factors, eight binary comparisons), there is only one instance
where the hypothetical application of a foreign characteristic to the U.S.
scene would *raise* U.S. energy consumption still further rather than dampening it, as happens typically. The situation arises in the case of Canada
where an even higher automotive energy intensity than prevails in the
United States is the principal explanation.

If one wishes to capture the essence of the exercise, the next-to-last
line of table 5-4, which compares the United States, on the one hand,
with the six-country Western European grouping, on the other, may be
worth a special glance. Per million dollars of GDP, U.S. utilization of
energy for passenger transport exceeds Western European by 134 tons
of oil equivalent. This difference, we may recall from table 5-1, explains
27 percent of the difference in the overall United States–Western European energy/GDP ratio. These 134 tons oil equivalent "decompose"
into the activity mix effect, followed closely by the role of intensity, with
the modal mix effect trailing relatively far back. Of course, the *combined*
mix effect, which is something of a synthetic construct, greatly outweighs
the intensity factor. Ignoring the interaction effects shown in table 5-4,
the 134 tons oil equivalent per million dollars of GDP difference between
the United States and Western Europe breaks down into the following
approximate percentages:

Due to activity mix effect	43 percent
Due to modal mix effect	18 percent
Due to intensity factors	39 percent

By reverting to table 5-3, we can better appreciate why these three elements break down into this order of importance. With respect to the
modal mix effect, it is true that compared to the United States, West
Europeans manage their travel activity with much greater proportionate
reliance on buses and trains (19 percent versus 3 percent). This is dramatic, in its way. But a converse comparison—the share of cars in total
passenger mileage—reveals a far less spectacular picture: the share of 92
percent in the United States and 80 percent in Western Europe, after all,
show the automotive mode to be dominant in both areas. By contrast,
activity mix (that is, passenger mileage relative to GDP) is 50 percent
greater in the United States than in Western Europe; while energy intensity of cars is nearly 60 percent higher in the United States.

To the extent that this statistical excursion conveys some element of policy significance, the following tentative thought is put forward. In terms of quantitative impact, adapting to the European activity mix (that is, to travel less relative to income) implies, in principle, the biggest pay-off for the United States. But size of country and other factors may impose limits on the extent of such an accommodation. On the other hand, if Europe is a guide, an alteration in the modal mix still leaves passenger transport dominated overwhelmingly by cars. This would point to reduced energy intensity for cars, as well as other modes,[7] as one of the more obvious potentials for enhanced energy conservation. This discussion leads more or less naturally to some thoughts—contained in the following section—of a more interpretative character.

Factors Underlying the Differences

By virtue of the fact that they have less travel activity, are more oriented to public transportation (which is itself characterized by relatively low energy intensity), and use cars that are less energy intensive, other countries use less energy for passenger transport relative to national output than does the United States. What predisposes other countries toward these more energy-sparing characteristics in the passenger transport field? Our discussion here is necessarily selective and tentative rather than exhaustive, systematic, and conclusive. Nonetheless, some fairly persuasive answers do suggest themselves.

RELATIVE PRICE. Relative price differences among countries unquestionably help shape the comparative transport patterns. The data in table 5-5, although covering only six of our nine countries, depict this aspect in a clear-cut fashion. As is frequently the case in this study, contrasts between the United States, the Western European group, and Japan are sharper than are the differences between the non–U.S. countries themselves, particularly within Western Europe.

We see in table 5-5, and most notably in the case of cars and gasoline and oil, the channeling of America's passenger transport expenditures (stated here in terms of shares of personal consumption expenditures)

---

[7] As table 5-3 shows, U.S. intensities tend to exceed those of other countries in cars, buses, and rail transport. The reason U.S. air travel shows up as less energy intensive is probably the fact that in domestic travel, which is what is covered here, the longer U.S. distances and other operating characteristics may promote greater fuel economies than are realizable in airline operations within small countries (for example, more taxiing, idling, takeoffs, and landings relative to cruising). Incidentally, the foreign aircraft intensity figures contained in our tabulations are exceedingly crude and should be regarded with a skeptical eye.

TABLE 5-5 Comparative Prices and Expenditures, Selected Passenger Transport Categories, Six Countries, 1970

(unit price is shown as index, U.S. = 100; quantity refers to amount purchased, as a percentage of personal consumption expenditures)

Transport categories	U.S.	France	W. Germany	Italy	U.K.	Japan
Cars						
Unit price	100.0	132.3	151.2	139.8	186.9	157.4
Quantity	4.2	1.6	1.6	1.4	1.3	0.3
Tires, tubes, accessories						
Unit price	100.0	138.1	151.2	108.7	78.7	128.3
Quantity	0.8	0.5	0.4	0.3	0.8	0.0
Repair charges						
Unit price	100.0	193.1	194.5	138.6	56.4	152.7
Quantity	1.2	0.8	0.6	0.5	1.1	0.2
Gasoline, oil						
Unit price	100.0	255.6	243.0	347.7	192.5	249.4
Quantity	3.4	0.7	1.1	0.6	1.1	0.2
Local public transport						
Unit price	100.0	73.6	77.6	62.7	70.2	26.6
Quantity	0.4	0.7	0.5	1.2	2.0	8.6
Rail transport						
Unit price	100.0	69.7	79.4	54.0	92.8	49.4
Quantity	0.0	0.6	1.1	0.9	0.4	1.4
Bus transport						
Unit price	100.0	113.2	79.7	72.7	76.4	62.0
Quantity	0.1	0.4	0.5	0.4	0.1	0.3
Air transport						
Unit price	100.0	143.3	143.3	140.3	128.9	106.8
Quantity	0.4	0.0	0.0	0.0	0.4	0.1
Parking, tolls, misc.						
Unit price	100.0	47.8	53.3	35.4	97.0	156.1
Quantity	0.7	1.6	1.1	1.5	1.1	0.2

Notes and Sources: Derived from ICP Report (Irving B. Kravis, Zoltan Kenessey, Alan Heston, and Robert Summers, A System of International Comparisons of Gross Product and Purchasing Power [Baltimore, published for the World Bank by The Johns Hopkins University Press, 1975]). The price relatives were obtained by deflating the purchasing-power-parity (ppp) exchange rate applicable to a given category by the ideal personal consumption expenditures (PCE) deflator. For example, for a unit of gasoline and oil, the ppp rate is 11.81 French francs per dollar; that is, it takes 11.81 francs to buy what a dollar buys. Deflating this 11.81 by the ideal rate of 4.62 yields the 255.6 price relative shown above. The quantities shown in the table reflect U.S. price weights. Other-country price weights might have altered the quantity shares somewhat but only for the United States–Japan comparison would the shares have been altered significantly. For example, compared to the U.S.–Japanese local public transport shares of 0.4 percent and 8.6 percent respectively (based on dollar price weights), the yen-weighted quantity shares come to 0.1 percent and 2.7 percent—a difference due to the vast price difference for this commodity.

Unlike other tables in this chapter which, in principle, exclude foreign travel, this table includes at least some components of foreign travel expenditures—air fares (as witness a U.K. percentage as high as the U.S. figure).

*FIGURE 5-1   Price Versus Quantity Purchased, Gasoline and Oil, and
Rail Travel, Six Countries, 1970*

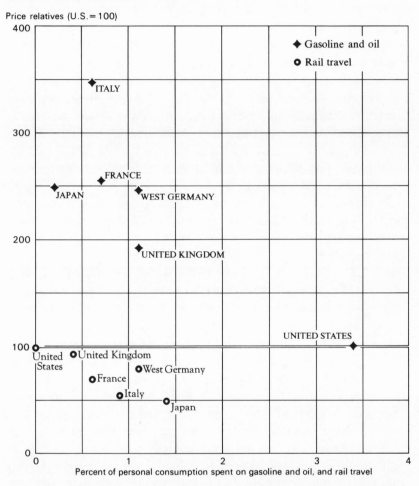

Percent of personal consumption spent on gasoline and oil, and rail travel

toward those categories in which the U.S. relative price is (by international standards) lowest. Table 5-5 discloses that in 1970 foreign motor
fuel prices ranged between 100 and 250 percent above levels prevailing
in the United States. Data in appendix F on international energy prices
for 1972 as well as information for 1975 indicate only a slight subsequent
shrinkage in this disparity. In a contrary fashion, foreigners' spending
for passenger transport is more oriented to those items—such as rail and
local public transport—which cost far less than in the United States.
Figure 5-1, which graphically illustrates the contrasting picture for gasoline and oil, on the one hand, and rail transport, on the other, is a

sufficiently crude approximation to the conventional demand curve to bear out this tendency. (The vertical and horizontal axes are proxies for, but not identical to, the usual denomination in terms of actual prices and actual quantities.)

More data points, both at a point in time, and over time, would be needed to establish the certitude of these relationships, let alone venture predictive judgments about demand elasticity and substitutability of travel modes. All we want to draw attention to is the unmistakable tendency for relative price differences for given transport modes between the United States and other countries to favor disproportionately greater automotive activity in the United States compared with other countries; and disproportionately greater public transport activity in other countries compared with the United States.

The data in table 5-5 do not allow us to compare the cost of, say, a "unit" of automotive travel (defined in terms of a standardized comparison of services rendered, such as passenger-miles) with a "unit" of rail travel, but rather permit us only to compare the cost of a given mode across the countries. But, in view of our earlier findings, we can safely infer that the high cost of foreign car ownership and operation constitutes a forceful reason for practicing economy in automotive energy use.

DENSITY AND SIZE. Although conscious of moving into a poorly charted behavioral territory, one is tempted to state two intuitive propositions linking density and size with energy use in passenger transportation. First, at comparable levels of per capita income, regions of high population density would seem to be more suited to public transport modes (and thus less energy-intensive travel) than would dispersed settlement patterns, the latter being relatively more attractive to automotive usage. Public transport systems, reflecting capital-intensive investments (subways being the most notorious example), require sufficiently high passenger load factors to meet these fixed costs as well as fuel, labor, and other charges so as to avoid either counterproductive high fares or degrees of public subsidization (which may or may not be forthcoming), or both. The requisite load factors are obviously more apt to be found in heavily built-up areas than in sparsely settled or dispersed ones. The presence and survival of a highly developed public transit network in a concentrated urban region like greater New York City and its absence in, say, the sprawling Los Angeles area, surely reflect this contrasting state of affairs.

Second, one would also guess that among equally prosperous countries, those with a large land mass would witness more travel activity

TABLE 5-6   Selected Data on Land Area, Population Density, and Passenger Transportation, 1972

Countries	Land area		Population density per square mile		Passenger-miles	
	Million square miles (1)	Square miles per $ million GDP (2)	Entire country (3)	Eight most populous urban areas (4)	Per $ thousand GDP (5)	Percentage by rail and bus (6)
United States	3.60	3.1	58	9,362	2,000	2.9
Canada	3.64	35.2	6	n.a.	1,386	7.8
France	0.21	1.0	245	13,431	954	20.4
W. Germany	0.10	0.4	643	12,709	1,472	17.6
Italy	0.12	0.8	468	n.a.	1,594	19.4
Netherlands	0.02	0.4	845	n.a.	1,256	18.7
United Kingdom	0.09	0.5	593	10,678	1,467	19.9
Sweden	0.17	4.2	47	n.a.	1,256	15.0
Japan	0.14	0.4	737	18,689	1,098	63.7

n.a. = not available.

Notes and Sources: Nationwide land area and density figures from United Nations, *Demographic Year-book 1973* (New York, UN, 1974). The eight-city density figures are based on unpublished, internationally standardized compilations developed and provided by the Foreign Demographic Analysis Division, Bureau of Economic Analysis, U.S. Department of Commerce. The urban density figures for the five countries shown represent simple arithmetic averages of eight urban areas in each of the countries. (A weighted average would assign too much weight to such places as greater New York City or Tokyo.)

of all kinds on a per capita basis or relative to GDP. When travel opportunities are limited by small areas, or where density brings the objectives of travel within close geographic reach, other activities take the place of travel (although traveling beyond the borders of one's country can obviously overcome these limits).

To what extent do the passenger transport characteristics of our nine countries fall more sensibly into place in the light of these density and size considerations? Unfortunately, the answer, on the strength of an admittedly cursory statistical reconnaissance, is at best suggestive. And of the two factors considered, density as an element in public transport appears to play the more demonstrable role.

Some statistics judged relevant to this issue appear in table 5-6. Sharp gradations in density do seem to accompany variability in the reliance on buses and railroads. Except for the Netherlands, Japan is the densest country and has by an overwhelming margin the largest share of public transport. The low-density-end of the spectrum is asymmetrical, however: the United States and Canada "fall into place" with relatively low proportions of public transport; but Sweden—also a low-density country, large parts of which are, like Canada, lightly populated—features an extensive public transport share.

Because bus and rail systems (keep in mind that the latter include subways and streetcars in our tables) predominate in and around cities, table 5-6 also shows available data on urban density for some countries. The largest urbanized areas of the United States are half as densely settled as those of Japan and between 70 and 90 percent as densely settled as those of the Western European countries shown. A more interesting phenomenon is not disclosed by the table. And that is the large number of *heavily populated* urban areas in the United States that are, in fact, characterized by low density. For example, the greater Dallas area just misses being on our list of the eight most populous U.S. urban areas (its 1972 population was 2 million), even though its population density is a relatively low 5,700 persons per square mile. By contrast, the seventeenth most populous French urban area (Nancy) had as high a density (13,000) as the eight-city French average. Another way of putting the same thing is to note that a populous city like Dallas can be less densely settled than many smaller population centers, such as Allentown, Pennsylvania.

These circumstances, which are recognized by students of urban geography but insufficiently treated in discussions of energy issues, seem to suggest that the kind of "critical mass" (in the form of high density and large population) needed for effective urban public transport may be much more prevalent outside North America. To deplore the fact that the 8.7-million-person Los Angeles urban area (with a density of 8,700 persons per square mile has deficient public transport while the smaller, yet more dense, city of Munich, Germany (1.6 million persons, 14,000 persons per square mile) has an effective one seems to us to ignore an important aspect of comparative history. Many American cities evolved in a twentieth-century, postautomotive period where a combination of abundant land, a new transport mode, and cheap fuels all pointed to unique patterns of living and transport. By contrast, the concentrated urban configuration of many European cities was firmly locked into place many years—if not centuries—earlier. It seems no accident that those American cities—namely the older ones along the Eastern seaboard, like Boston, New York, and Philadelphia—which most closely resemble European cities are also the ones in which public transportation survives as an enduring tradition.

Finally, what might emerge from the statistics as a beneficent mix of cars and public transport might, in the visual context of a crowded, polluted metropolis, appear in an altogether less favorable light. Take Japan, with its dramatically low incidence of car ownership: few would point with pride at Tokyo's incredibly car-congested roads. And indeed, the number of cars relative to square miles of land area is three times higher in Japan than in the United States.

However, our proposition regarding travel propensities in large and small countries meets tougher empirical resistance. At least, our nine-country sample is inadequate for a thorough exploration of the question. The second and fifth columns of table 5-6 show no telling evidence that, relative to income, a lot of land means a lot of travel. Both Canada and Sweden exceed the U.S. land/GDP ratio, yet each falls below the U.S. passenger-mile/GDP ratio. One might perhaps argue that much of both Canada's and Sweden's territory is really inaccessible to normal travel (both countries possess vast and sparsely settled Arctic regions) and that such an allowance would produce a picture more congenial to our provisional judgment.[8] For example, a handful of dense, intercity corridors, such as Stockholm-Malmo or Stockholm-Gothenberg, hold perhaps 90 percent of Sweden's population. But fruitful research on this score necessitates a degree of effort that goes well beyond the scope of our study.

TRENDS OVER TIME. A "snapshot" view based on one year risks missing potentially important year-to-year changes among the countries. If such changes could be coupled to what is essentially a static analysis for 1972, we might well be prompted to modify our interpretation of significant intercountry differences. One could argue that certain features of an economy—for example, industrial structure, manufacturing processes, or housing characteristics—change sufficiently slowly as not to violate the use of cross-sectional data for reaching conclusions with some enduring validity. In the field of passenger transportation, however, it is possible that significant questions are cloaked by the cross-sectional approach.

It is apparent that the automobile has played a major role in shaping the present U.S. style of life and the structure of its metropolitan areas. It is also apparent, albeit with a time lag involved, that the automobile is having a similar effect on the rest of the world. It is this lag which a cross-sectional comparison by itself fails to capture. Table 5-7 provides unmistakable evidence of a convergence toward the United States in passenger car diffusion (measured here by comparative data on car ownership per

---

[8] We attempted a crude effort along these lines by constructing an "adjusted" land/GDP ratio, in which the adjustment factors reflected the country's density. The idea being that: what good is a lot of land if devoid of inhabitants? The results proved chaotic.

TABLE 5-7 Ownership of Passenger Cars Relative to Population, Selected Years Between 1953 and 1974

Years	U.S.	Canada	France	W. Germany	Italy	Netherlands	U.K.	Sweden	Japan
1953									
Cars per 1,000 persons	289	169	47	24	13	18	55	60	1
Index, U.S. = 100	100	58	16	8	4	6	19	21	0.3
1961									
Cars per 1,000 persons	344	237	133	92	48	53	113	173	7
Index, U.S. = 100	100	69	37	27	14	15	33	50	2
1965									
Cars per 1,000 persons	386	269	197	153	105	103	166	232	22
Index, U.S. = 100	100	70	51	40	27	27	43	60	6
1972									
Cars per 1,000 persons	462	319	269	253	229	229	230	303	119
Index, U.S. = 100	100	69	58	55	50	50	50	66	26
1974									
Cars per 1,000 persons	492	377	269	272	258	254	246	323	144
Index, U.S. = 100	100	77	58	55	52	52	50	66	29

Source: United Nations, *Statistical Yearbook* (New York, UN, various issues).

thousand inhabitants). Thus, as recently as 1961, the car/population ratio in Holland was 15 percent that of the United States; by 1972, it had moved to within 50 percent. For Britain, the ratio went from 33 to 50 percent. (Per capita car ownership in the United States was itself rising by nearly 3 percent per year during this period.) As the table shows, the trend for Japan and Italy was even more spectacularly upward. The spread in per capita car ownership was shrinking much faster than the spread in real per capita income.

Few countries seem to be immune to the spread of suburbanization. Along with it goes people's insistence on the kind of mobility that only automobiles are deemed capable of satisfying. It is interesting to note, for example, that even in Holland, the volume of passenger-miles represented by public transportation systems remained virtually stagnant between 1955 and 1972, while automotive use increased by more than 10 percent per year. During the same period, the proportion of public transport in total passenger-miles changed from one-half to less than one-fifth.[9]

As in Holland, so in the United Kingdom: bus and rail patronage declined in absolute terms during the 1962–72 decade (by about 2 percent yearly), while automotive use rose 7½ percent yearly.[10] The fact that British rail energy intensity (table 5-3) exceeds that of U.S. passenger railways attests to the low U.K. load factor—a situation which contemplated British policy seeks to "ameliorate"—by abandonment of a large volume of trackage. Even though countries like Britain, Holland, France, and (especially) Japan continue to depend, relatively speaking, far more on public transport than the United States, the fact that these countries' car population is frequently imposed on smaller land areas can produce urban automotive congestion which need not put the United States to shame.

We do not argue the irreversibility of such trends, which, incidentally, have also included a gradual increase in the size of foreign cars relative to the size of U.S. cars. Technological developments, public policies, real price increases, and independent shifts in consumer preferences may all bring about new—and, from an energy point of view, more benign—arrangements in passenger transportation. For example, although the *level* of foreign automotive fuel prices has for some time stood substantially above that in the United States, an end to the stable or even real

[9] J. A. Over, ed., *Energy Conservation: Ways and Means* (The Hague, Future Shape of Technology Foundation, 1974) p. 113.

[10] U.K. National Economic Development Office, *Energy Conservation in the United Kingdom* (London, HMSO, 1974) p. 41.

price declines that characterized the pre-1973 period could trigger dampened automotive reliance in the future. For the present, however, it is hard to avoid the impression of some considerable multicountry convergence toward a preeminent role for the automobile as typified by the United States.

PUBLIC POLICY MEASURES. Public policies—taxes, subsidies, expenditure programs, and regulatory measures—can exercise a "carrot-and-stick" role in the determination of national transportation patterns. Of course, deterrent and incentive devices, being very much creatures of political processes and institutions, have the potential not only for promoting rational objectives, but also for distorting economic and social priorities, however these value-laden terms may be defined. In any case, it should be clear that at least some of the energy transport characteristics which we have touched upon are manifestations of public policy making.

An obvious example of public intervention in passenger transport energy markets is the imposition of excise taxes on gasoline. These taxes typically have been many times steeper overseas than they have been in the United States and Canada and are virtually the entire reason for the differential automotive fuel prices that we observed earlier. These price differentials, in turn, were seen to be associated markedly with variability in motor fuel consumption levels. It is worth noting, as an aside, that heavy foreign taxation of gasoline had its genesis largely in the quest for governmental revenues rather than in enlightened transportation control strategies. Such unplanned benefits as discouraging acquisition of cars with poor gasoline economy and making public transport a more palatable alternative have ensued. (In the wake of the 1973–1974 oil crisis, a more purposeful resort to gasoline taxes as a conservation measure was taken in a number of countries.)

In addition to automotive fuel taxes, overseas governments typically levy much higher registration fees and sales taxes on car ownership and acquisition. Ownership of what would, by American standards, be regarded as a small car has, in recent years, cost a French person over 2,000 francs (about $450) annually in taxes.[11] Value-added taxes add 33⅓ percent to the purchase price. France, along with Italy, seems to be at the high end of the tax range among our countries. But levies elsewhere take a hefty bite, judged by American taxes, and even more so in relation to foreign incomes. For example, the West German Motor Vehicle Tax Law of 1972 specified an annual vehicle tax equivalent to

11 Tax rates rise steeply with horsepower but decline with the age of the car.

about $8.80 per 200 kg or $50 for a 2,500-pound automobile. On top of that, there is an annual license fee amounting to about $150. In Sweden, too, substantial fuel and weight taxes encourage ownership of light cars, to which the paucity of cars weighing above 3,600 pounds (the U.S. average) attests.

Complementary to policies determining private automotive activity are government measures designed to sustain the viability of public transport systems. Passenger railways, in particular, are heavily subsidized throughout the world. And where a rail system flourishes, this usually means that subsidies flourish to accommodate large losses. And where, as in Britain, such government subvention nonetheless fails to attract adequate ridership, such losses and subsidies spiral while, as noted before, passenger mileage stagnates.

The point here is certainly not to inveigh against the desirability of subsidized public transport systems. On the contrary, as we shall note in a moment, the social utility of such transport arrangements may often transcend accounting losses. Rather, it is to lead up to the recognition that a transport energy comparison whose "bottom line" revolves around intercountry energy intensities (often labeled, somewhat unfortunately, "energy efficiencies") may be a seriously deficient analysis. For it leaves out the nonenergy cost (labor, investment capital, other resources) of making a good energy/output "showing." It may well be that if higher overall resource costs *are* encountered in moving from high to low energy intensities, these might easily be justified by social benefit–cost criteria built around such things as lessened congestion, air pollution, oil import dependence, noise, and other disamenities of automotive use. But without a reasonably explicit indication of the magnitude of such trade-offs, an analysis consisting primarily of energy dimensions constitutes at best a useful but only partial explanation of important questions. The discussion of energy and passenger transportation in this chapter—as is the case throughout our entire study—represents just such a limited treatment. As such, it is designed to offer some worthwhile insight; but as our reference to public policy measures demonstrates, decisive issues of *overall* resource use are still with us.

Freight Transport

A paucity of comparative international data on energy and freight transport permits making only some cursory observations in this area. Moreover, our review is confined to a comparison of the United States with the grouping of six Western European countries. However, what summary data we have been able to assemble disclose some striking comparative energy characteristics of freight transport.

Table 5-8, arranged in a format similar to the passenger transport data shown earlier in table 5-3, identifies the key features of freight energy consumption in the United States and Western Europe: the amounts of freight traffic relative to the size of the respective economics (expressed as ton-miles per dollar of GDP); the modal mix of ton-miles —distributed among rail, trucks, waterborne commerce, air, and pipelines; and the energy intensity associated with each of these modes. Unfortunately, a modal breakdown of intensities was available only for the United States; for the Western European group, we must be content with an aggregated intensity for all modes combined.

Note the contrasting modal mix pattern. While both regions depend on rail for about a third of their freight traffic, the other modal shares are distinctly different. The six European countries are much more heavily wedded to trucking rather than the United States; conversely, U.S. freight consists disproportionately of waterborne modes and pipelines.[12] If the U.S. modal intensities shown in the table are a reasonably fair representation of what is likely to be the case abroad, this means that U.S. freight traffic is weighted toward less energy intensity than is Western Europe. The weighted total intensities for the two areas bear this out: energy consumption per million ton-miles is nearly twice as high in Europe as it is in the United States. Indeed, the figures suggest that one or more individual European modes may be more energy intensive than in the United States. For, when we apply the individual U.S. intensities to the respective European ton-miles, we derive a weighted European overall intensity of only 50 tons of oil equivalent per million ton-miles rather than the actual estimated figure of 57 tons of oil equivalent—in other words a hypothetical 70 percent higher than the U.S. intensity of 29 rather than the 100 percent higher level that actually prevails.

The other striking fact shown in table 5-8 is the multiple of four by which the United States exceeds the six-nation Western European group in volume of freight traffic per unit of GDP. No doubt, an important

---

12 Energy used for pipeline movement (that is, for pumping purposes) is legitimately a freight energy component, although its treatment in the basic OECD statistics is ambiguous. If, as seems likely, pipeline energy use is implicitly classed in the "energy" rather than "transport" sector in the OECD-based data in chapter 3, our inclusion in the present chapter of pipelines within a correspondingly understated *total* transport energy use (passenger, freight, other) creates a very slight degree of statistical distortion.

TABLE 5-8 Freight Transportation: Ton-miles, Energy Consumption, and Related Data, by Transport Mode, 1972

| Transport mode | Energy consumption | | Ton-miles | | Energy consumption (tons oil equiv. per million ton-miles) | Ton-miles per dollar GDP | Energy/GDP ratio (tons oil equiv. per $ million GDP) |
	Million tons oil equiv.	Percentage	Millions	Percentage			
United States							
Rail	13.9	21.7	713,000	32.6	19.5	0.61	12
Trucks	37.9	59.1	419,309	19.1	90.4	0.36	32
Water	7.5	11.7	573,707	26.2	13.1	0.49	6
Air	2.725	4.2	3,353	0.2	812.7	0.00	2
Pipeline	2.15	3.4	480,760	22.0	4.5	0.41	2
Total	64.175	100.0	2,190,129	100.0	29.3	1.86	54
Six Western European countries							
Rail	n.a.	n.a.	121,002	30.1	n.a.	0.14	n.a.
Trucks	n.a.	n.a.	186,114	46.2	n.a.	0.21	n.a.
Water	n.a.	n.a.	58,721	14.6	n.a.	0.07	n.a.
Air	n.a.	n.a.	72	0.0	n.a.	0.00	n.a.
Pipeline	n.a.	n.a.	36,591	9.1	n.a.	0.04	n.a.
Total	23.11	100.0	402,500	100.0	57.4	0.46	26

Note: n.a. = Not available.

Sources: The U.S. energy consumption and ton-mile data are taken from Federal Energy Administration, *Project Independence and Energy Conservation: Transportation Sectors*, final task force report (Washington, D.C., GPO, 1974). European ton-mile figures came largely from United Nations Economic Commission for Europe, *Annual Bulletin of Transport Statistics, 1972* (Geneva, ECE, 1973). The pipeline component had to be added separately and represents a rough estimate based on spotty data from a number of national sources. The European energy consumption figure is also an approximation, derived as follows: an upper limit to freight energy consumption is 42.02 million tons oil equivalent, representing the freight and other category shown in table 5-1. For the United States, freight represents half of "freight and other." We assume that this one-half represents an absolute minimum European share, since the "other" portion of "freight and other" is dominated by such things as school buses and general aviation, which can be taken to be relatively more important in the United States. Not wishing to overestimate European freight consumption (thus implying a higher intensity than warranted), we assumed, conservatively, we believe, that freight might reasonably be estimated at 55 percent of the "freight and other" total. The reader should keep in mind that, in principle, all data here presented refer to *domestic* transportation within the countries tabulated. (That is, not even the intra-six-country freight traffic—say, between Holland and Germany—is included.) This means that the addition of international freight traffic would serve to increase European ton-miles and freight energy consumption proportionately much more than those of the United States.

underlying factor in this contrasting picture is the fact that the land area of the United States is more than five times the combined land area of the six European nations, while the U.S. GDP is only 1.3 times as large. It would seem that a given amount of economic activity simply involves a lengthier geographic spread in the production–consumption chain in the United States. (Long distance haulage of such export commodities as grains and coal would contribute to U.S. energy consumption even though essentially a consequence of overseas economic activity.) If this is a fitting interpretation for freight traffic, it is one which we found to be not particularly suitable in analogous treatment of passenger transport; that is, size of country by itself was seen to be a deficient explanation for variations in the activity mix effect.

While geographic spread and length of journey impose a disproportionately large volume of freight traffic (relative to GDP) upon the United States, the same phenomenon probably acts to favor an energy-conserving modal mix. For these same long distances are economically more amenable to rail haulage than would be a more distinctly short-haul pattern of freight activity.

It is worth remembering once again that in principle we exclude external freight traffic, although the underlying data are sometimes obscure as to coverage. To the extent that foreign traffic is indeed omitted, its addition would raise European ton-miles by a relatively greater amount than U.S. freight. This would have the effect of dampening the United States–Western European differential in the freight energy/GDP ratio shown in the last column of table 5-8—a differential which, as we could anticipate from what has already been said but can see more clearly from table 5-9, is totally dominated by the activity mix.

Table 5-9 is analogous to the passenger transport analysis in table 5-4. It shows how the effect of both the differing U.S. modal composition as well as the smaller freight energy intensity act to shrink the freight energy/GDP difference, only to be more than counterbalanced by the greater number of U.S. ton-miles generated relative to GDP. Even if energy intensity were synonymous with energy efficiency—and our position is that this cannot be assumed—here is a case where an intercountry spread in the energy/GDP ratio is surely not an obvious reflection of an efficiency gap; indeed the energy/GDP differential may obscure U.S. "superiority" in energy freight efficiency compared to the Western European group.

We have to be content here with having sketched out a few comparative characteristics of energy consumption in freight transport. As in the

TABLE 5-9   Analytical Breakdown of Differences Between United States and Six Western European Countries in Freight Transport Energy/GDP Ratios, 1972

(tons oil equivalent per $ million GDP)

Due to mix factors				Due to intensity factors and to interaction between total mix and intensity	Total actual difference (U.S. minus 6 countries)
Modal mix effect	Activity mix effect	Interaction	Total mix effect		
−35	40	26	31	−3	28

Notes and sources: Based on data shown in table 5-8 and derived as explained in appendix E, with the following exceptions: we possessed no data on the energy intensity of individual European freight transport modes, but were able to construct only an overall European freight intensity estimate. Consequently, we were not able to hypothesize the effect of applying European intensities to the different U.S. modes. (In the case of passenger transport, this effect is captured in the fifth column of table 5-4.) Thus, the fifth column above is derived as a residual and reflects the effect of both intensity and interaction. This problem does not impair the major significance of the table, as discussed in the accompanying text.

case of passenger transportation—perhaps even more so—a sensible discussion of the extent to which the movement of goods departs from some ideal economic and energy yardstick would involve us in formidable technical and policy questions. Suffice it to say, patterns of freight transport, as pictured above, reflect a mixture of economic logic and public policy measures—the latter being of both the "special interest" and enlightened varieties. For example, desirable as it may appear to be from an energy point of view, a shift from road to rail of short-haul traffic (say, below 100 miles) is probably an untenable proposition in most countries. Conversely, long-haul movement of bulk commodities is already performed to a large extent by rail.[13] But a country's geography and industrial patterns will clearly also affect modal choices and how they perform.

At the same time, the existing freight transport energy features—as pictured in our statistics—need not be taken as reflecting the ultimate in logic. People everywhere are concerned about the environmental consequences of the growth of road transport, and whether taxes and other policy instruments adequately cover real resource costs (such as roadways) as well as environmental impacts is clearly an unsettled question.

In the United States, there has been much discussion of whether the extensive regulatory controls exercised by the Interstate Commerce Com-

[13] For example, if in Britain "*all* freight movements of more than 100 miles were transferred to rail, total road traffic would be reduced by only 2–4 percent." (*Transport Policy,* Volume 1 [London: HMSO, 1976, pp. 14–15].)

mission over trucking (for example, with respect to entry of firms, rates, routes, and backhauls) has contributed to a warped modal mix inimical to railroad freight or to an unnecessarily high energy intensity in trucking.[14] But, here again, the issues are intricate and unresolved. To take only a minor example: a commitment to serve out-of-the-way communities with small cargoes and potentially "empty" mileage on the backhaul can look unfavorable in an energy sense though defensible in other respects.

Summary

The principal findings of this chapter can be summarized as follows:

1. Among sectors separately identified in this study, transport is a major explanation for energy/GDP variability between the United States and most of the other countries surveyed. For example, it accounts for 40 percent of the energy/GDP difference between the United States and Western Europe (treated as a whole).

2. Within transport, both passenger activity and freight contribute to this difference, but the passenger component tends to be much more important.

3. In the case of passenger transport, we quantified three contributing elements to energy/GDP variability: the modal mix (the proportionate share in total passenger miles of different forms of transport); the activity mix (the relationship between travel activity in passenger-miles and GDP); and the energy intensity of given modes (that is, energy consumed per passenger-mile).

4. Each of these three elements enters into an accounting of energy/GDP differences. But their proportionate importance varies, as can be seen from a comparison—simplified for purposes of this summary—of the United States with the six-nation Western European group:

Percentage of energy/GDP difference due to:

activity mix effect	43 percent
modal mix effect	18 percent
intensity factors	39 percent

[14] As in the case of passenger transportation data, the aggregative nature of the freight statistics obscures a number of comparative insights. For example, in a comparison of freight movement within *just urban areas,* Sweden displays a lower overall energy intensity than the United States.

5. Relative prices of car operation and ownership, on the one hand, and of public transport, on the other, are associated quite markedly with the relative use of these respective modes, although the contrast is more vivid when looking at the United States alongside a group of other countries, rather than among third countries.

6. In turn, these relative price differences reflect tax and other public policy measures, which in a number of places outside the United States are congenial to public transport and constitute a deterrent—at least to an extent—to private motoring. For example, in 1972 automotive fuel elsewhere was typically taxed to the extent that the prices were two to three times those in the United States.

7. The degree to which these observed price–quantity relationships constitute the basis for North American policy approaches that would use price to inhibit automotive use is problematical. One cannot dismiss the findings, however limited, as insignificant. But whether they constitute a widely applicable elasticity measure (let alone whether they are politically feasible) is another question. To name just one separate factor that appears to represent a noticeable influence on comparative transport patterns: characteristically high urban densities of Europe and Japan, in contrast to the dispersed North American settlement pattern, surely contribute in an important way to a transport split more balanced toward public modes of travel. However, even this more energy-saving modal pattern in other countries, when coupled to a limited land area, does not preclude automotive congestion similar to or greater than that observed in U.S. urban areas. (Some may prefer not to cite density as a "separate" factor from price, holding that settlement patterns are themselves partially a price-induced phenomenon. But the practicability of shaping or reversing such locational characteristics through energy price and other policy instruments, much less hypothesizing their net energy consequences, is a topic beyond our purview.)

8. Having said all that, let us recall that, outside Japan, automobiles are the overwhelming form of passenger transport in *all* countries and that automotive displacement of alternative modes has been proceeding dramatically during the past ten to twenty years.

9. Freight transport also contributes to the higher U.S. energy/GDP ratio. Most interestingly, however, this comes about exclusively by virtue of the high volume of traffic (relative to GDP) that is generated in the United States compared to the grouping of European

countries that we analyzed. Indeed, the U.S. freight modal mix is, more than Western Europe's, oriented to such energy-saving forms as rail, pipelines, and waterborne traffic. If one argues that size of country and long-distance haulage of bulk commodities (such as ores, grains, and coal) are inherent characteristics of the U.S. economic structure and geography, there is here a case where a relatively high energy/GDP ratio is in no obvious way reflective of comparative energy "inefficiency."

We conclude that the transport sector offers exceedingly useful insights into the anatomy of intercountry energy/GDP variability. There undoubtedly are some significant lessons to be pondered and to be applied; for example, concerning a role for less energy intensive vehicles and an enduring place for public transport forms in energy conservation strategies. But it is also necessary to recognize that transport characteristics reflect a number of deeply imbedded economic, geographic, and social features whose purposeful change, in the interest of energy savings, can be contemplated only within a much wider framework of costs, resources, and social preferences than we have fashioned in this chapter or this study.

CHAPTER 6

Industrial Sector

An Aggregate Perspective

IN THIS CHAPTER we continue the discussion of comparative sectoral patterns of energy consumption begun in chapter 4. The topic is energy consumed in industry. It should be noted that for purposes of this chapter the energy sector (that is, extraction, refining, and electric and gas utilities) is combined with industry.[1]

Industry Within Total Energy Consumption

Chapter 3 pointed to the important position of industry among the other energy-using sectors of the economy identified in this study. To recapitulate briefly from the findings discussed in chapter 3: industry is a leading source of energy demand in our nine countries; it is quantitatively the most important energy-consuming sector, outranking the household-commercial sector in Italy, the United Kingdom, and Japan; it is essentially on the same level of importance as the household-commercial sector in France and Germany, while elsewhere it ranks second—trailing the household-commercial sector; and when the definition of industry is broadened to include the energy sector itself, it outranks other energy-using sectors of the economy in all of our countries except Holland, where it remains in second place. Table 6-1, compiled from our basic statistical tables in appendix B, contains the data in question. The virtue of proceeding with the discussion using the broadened industrial category, which includes the energy sector shown in the second column of table 6-1, as our total is that this permits us to relate energy relationships to a similarly defined sectoral component of GDP, as we shall note a bit further on.

One figure disclosed in table 6-1—the rather low proportion of nationwide energy use accounted for by Sweden's energy-producing sector—is

---

[1] Transformation losses, however, are dealt with in a separate section at the end of the chapter, and nonenergy uses of energy commodities (for example, road oil) are excluded altogether.

TABLE 6-1   Energy Consumed by the Industrial Sector as a Percentage of Total Energy Consumption,[a] *1972*

Countries	Excluding energy sector	Energy sector	Including energy sector
U.S.	20.9	9.1	30.0
Canada	21.9	7.2	29.1
France	27.5	7.2	34.7
W. Germany	29.0	7.2	36.2
Italy	30.8	5.2	36.0
Netherlands	20.0	7.8	27.8
U.K.	28.4	7.2	35.6
Sweden	25.9	3.1	29.0
Japan	38.9	5.7	44.6

Note: Energy sector refers to energy consumption by energy producers (for example, coal mining, natural gas, and crude oil extraction) and processors (for example, refineries and utilities). Does not include transformation losses incurred in converting one energy form into another. But electricity and gas distribution losses are included.
Source: Data in appendix B.
[a] Total inland consumption as defined by OECD.

worth noting. It appears to stem from several factors. By comparison with other countries on our list, Sweden has relatively small sectors devoted to energy extraction and petroleum refining, and Sweden's predominantly hydroelectric generating stations are estimated by OECD to absorb less internal electricity in their operation than the fossil-fueled facilities that predominate in other countries. This characteristic seems to more than compensate for the disproportionately greater use of electricity as an energy form in Sweden.

This last point is revealed in table 6-2, also taken from data in appendix B. It shows the proportion of energy used by industry and the energy sector that is consumed in the form of electricity:

TABLE 6-2   Electricity as a Percentage of Energy Consumed by the Industrial Sector, 1972

Countries	Excluding energy sector	Energy sector	Including energy sector
U.S.	18.0	15.8	17.3
Canada	23.0	18.4	21.8
France	14.4	17.8	15.1
W. Germany	14.6	21.6	16.0
Italy	15.5	26.1	17.1
Netherlands	14.8	11.7	13.9
U.K.	11.8	26.5	14.8
Sweden	28.5	56.3	31.5
Japan	19.3	25.0	20.0

The thing to notice in this array is the fact that the United States does not exhibit a great propensity to use electricity in its industrial operations in comparison with the other countries. For example, in Sweden and Canada—the two countries rich in hydroelectricity—where there is a lower ratio than in the United States of electricity-to-thermal energy costs, the tabulated electricity shares are very high.

The final bit of recapitulation from chapter 3—which brings us back to the major focus of the study—has to do with the impact of industry on overall energy/GDP ratios. Table 6-3 presents pertinent data. The intercountry industrial energy/GDP variability pictured in column 6 of the table spans a considerable range relative to the United States. Canada's industrial energy/GDP ratio exceeds the U.S. figure—which is the second highest among the nine countries—by 16 percent. The ratios for the other seven countries trail the United States by 10 to 38 percent, the simple average working out to a bit more than 20 percent below the United States. (Here we are talking—and will be for the balance of the discussion in this section—about industry in its broadened scope, which includes energy-producing industries.)

Reasons for Variability in Industrial Energy Use

This dispersion in the industrial energy/GDP ratio, while marked, tends nonetheless to be narrower than that in other energy-using sectors. Thus, transport and households fall further below the United States in their respective sectoral ratios than does industry. The proportion of the overall energy/GDP variability that originates in the industrial sector reflects the same phenomenon. As column (9) of table 6-3 shows, the United Kingdom's industrial sector accounts for 13 percent of the overall energy/GDP gap between that nation and the United States. Yet the transport and household-commercial sectors account for about 50 and 30 percent respectively. An identical pattern of dispersion is not duplicated in each of our pairs of comparisons with the United States. Indeed, there is little that ever falls into place in an uncomplicated fashion. Still, it is a fair generalization that variability in the overall energy/GDP ratio would be much smaller if it were primarily a manifestation of *industrial* energy consumption (relative to GDP), and thus, there would probably be far less fascination with energy/GDP relationships.

One might be tempted to connect that observation with the following type of intuitive reasoning. After all, we are dealing for the most part here with a group of highly advanced economies—a condition which suggests roughly comparable, if not identical, degrees of industrial pene-

TABLE 6-3 *Energy Consumption in the Industrial Sector Relative to GDP, 1972*

Countries	Excluding energy sector (1)	Energy sector (2)	Including energy sector (3)	Excluding energy sector (4)	Energy sector (5)	Including energy sector (6)	Excluding energy sector (7)	Energy sector (8)	Including energy sector (9)
	(tons oil equiv. per $ million GDP)			(———index, U.S. = 100———)			Percentage difference in energy/GDP ratios between U.S. and other countries accounted for by industrial energy consumption[a]		
U.S.	309	135	444	100.0	100.0	100.0	—	—	—
Canada [a]	388	129	517	125.6	95.6	116.4	27.1	-2.1	25.0
France	219	57	276	70.9	42.2	62.1	13.1	11.4	24.5
W. Germany	299	74	373	96.8	54.8	84.0	2.2	13.6	15.8
Italy	282	48	330	91.3	35.6	74.3	4.8	15.4	20.2
Netherlands	254	100	354	82.2	74.1	79.7	26.4	16.8	43.2
U.K.	318	81	399	102.9	60.0	89.9	-2.6	15.5	12.9
Sweden	275	33	308	89.0	24.4	69.4	8.1	24.4	32.5
Japan	330	48	378	106.8	35.6	85.1	-3.3	13.8	10.5

Note: Blanks = not applicable.
Source: Data in appendix B.

[a] A positive percentage in the last three columns signifies, except for Canada, what portion of the difference between the U.S. aggregate energy/GDP ratio and the other country's energy/GDP ratio is attributable to the sectors shown here. A minus sign means that, although the United States has a higher overall energy/GDP ratio, the sectoral ratio is higher for the country shown. In the case of Canada—whose overall energy/GDP ratio exceeds that of the United States—a positive percentage describes the share of the overall Canadian excess originating in a given sector.

tration, and with countries none of which is denied the energy-using manufacturing processes and technologies that flourish elsewhere. This suggests, perhaps, a band within which no significant differences in energy intensities prevail.

The comparative ratios of industrial energy consumption to total GDP described in this chapter are not out of line with such a mind-set. In fact, however, it turns out that far more intricate elements are at work here which need to be sorted out. As a first necessary—although, admittedly, far from sufficient—step, what we have tried to do is to decompose industrial energy use, relative to GDP, into (a) the portion attributable to "structure"—that is, the share of GDP originating in the industrial sector; and (b) the portion attributable to energy "intensities"—that is, the amount of energy associated with each unit of GDP generated in the industrial sector of the economy. In a way, the treatment is analogous to the approach devised at various other points in this study; for example, in the preceding chapter's discussion of transport. Note that structure, in this context, refers to the share of industry in the economy as a whole, not to the internal composition of the industrial sector. *Intra*-industrial aspects will be discussed under Industry Profiles later in this chapter.

There are a number of caveats surrounding this exercise that are too important to be left to footnotes or appendixes, for they temper—in a very significant way—the certainty of any interpretation placed upon the statistical findings. The most important of these provisos are discussed below.

1. In order to compare the association of sectoral energy use and sectoral economic activity of our nine countries, it is necessary to use standardized data on GDP originating in the industrial sector. However, we must admit to (a) a degree of unease about whether for a given source the sectoral definitions of GDP are truly comparable from one country to another, and (b) even greater concern over the extent to which a definition of a sector in one source meshes with the definition of ostensibly the same sector elsewhere. For estimates of GDP originating in industry, we relied largely on the OECD *National Accounts of OECD Countries* and, secondarily, on the *UN Yearbook of National Accounts Statistics*.[2] The second of these concerns emerged particularly in the course of fashioning UN statistics into the controlling OECD mold.

[2] Organisation for Economic Co-operation and Development, *National Accounts of OECD Countries, 1961–73* (Paris, OECD, 1974); United Nations, *UN Yearbook of National Accounts Statistics* (New York, various years).

2. While the availability of the ICP study[3] permitted the use of purchasing-power-parity (PPP) exchange rates for the derivation of GDPs in dollars, we have no intercountry relative price data enabling us to estimate the industrial component of GDP on a similarly "real" basis of comparison. Having had to fall back on use of the economy-wide "ideal" PPP exchange rates, we are consequently in no position to assess the reliability of the constant-dollar comparability of the ensuing industrial GDP estimates. In contrast, the treatment of the iron and steel industry later in this chapter benefits from the use of reasonably homogeneous physical output measures—tons of steel ingots—rather than the dubious dollar comparisons.

3. It was necessary to achieve as much comparability as possible between the energy-using components of industrial energy consumption, on the one hand, and the composition of industrial GDP, on the other. Thus, by broadening the scope of industrial GDP to include manufacturing, mining, and construction (which comprise the industrial sector as defined in energy consumption statistics), a fair degree of comparability seems assured. We do know of one element of mismatch: industrial energy usage in this study includes (in its broadened definition) consumption within electric and gas utilities, along with such other, and more important, energy industries as petroleum refineries. Unfortunately, industrial GDP could not be augmented to reflect GDP originating in utilities because the national accounts data combine local electric and gas utilities along with water and sanitary services, and there was no way to disentangle those specific parts. However, since the amount of energy used by electric and gas utilities themselves is rather modest, inclusion in industrial energy consumption and exclusion in the corresponding GDP component is deemed a tolerable inconsistency.

4. Industrial use of wastes as an energy source (for example, utilization of wood wastes in Sweden) could not be counted because there are no systematic and comparable data available for the nine countries. Also, in the case of co-generation of electricity and steam, use of the OECD energy format, which fails to identify the use of energy in the form of steam, produces some underestimation in industrial consumption of energy. This omission may blemish U.S. comparisons with individual countries, but does not distort the overall picture.

[3] Irving B. Kravis, Zoltan Kenessey, Alan Heston, and Robert A. Summers, *A System of International Comparisons of Gross Product and Purchasing Power* (Baltimore, published for the World Bank by The Johns Hopkins University Press, 1975).

TABLE 6-4   Energy Consumption in Industry, and GDP Originating in Industry, 1972

Countries	GDP originating in industry as percentage of national GDP	Industrial energy consumption		Difference from U.S. in ratio of industrial energy consumption to GDP (tons oil equiv. per $ million GDP)
		Tons oil equiv. per $ million industrial GDP	Tons oil equiv. per $ million national GDP	
U.S.	31.1	1,427.7	443.7	—
Canada	29.0	1,777.2	515.7	−72.0
France	48.1	574.0	275.9	167.8
W. Germany	50.7	736.6	373.5	70.2
Italy	33.3	990.5	329.6	114.1
Netherlands	35.6	995.4	354.1	89.6
U.K.	34.8	1,146.1	399.1	44.6
Sweden	25.7	1,198.1	308.0	135.7
Six Western European countries	41.8	822.1	344.0	99.7
Japan	40.9	924.6	378.5	65.2

Note: The concept of GDP originating in industry is similar to, but not identical with, value added by industry. See accompanying text for further remarks concerning reliability of these estimates. Blank = not applicable.

Sources: Energy consumption data and total GDP based on appendixes B and A respectively. "GDP originating in industry" derived from various issues of *UN Yearbook of National Accounts Statistics* and OECD, *National Accounts of OECD Countries.*

5. Finally, the figures relating total industrial energy use to total industrial GDP, no matter what they reveal, must be judged in the light of their highly aggregative nature. The degree of aggregation is such as to very likely conceal the *intra*-industrial mix and energy-intensity characteristics (or both) that, ideally, would enter into comparative international analysis. This aspect really shifts us over to the question of findings and their significance, so it is a good point at which to proceed with the substantive discussion.

The data in table 6-4 show the nine-country variability in both the proportionate importance of industry in the economy (as measured by share of national GDP originating in industry) and the ratio of industrial energy consumption relative to industrial GDP—that is, industrial energy intensity. The United States is near the bottom of the range of industrial GDP shares; only Canada and Sweden are lower. Conversely, the United States is near the top of industrial energy intensities, being exceeded only by Canada. Just these simple observations suffice to point to the principal

TABLE 6-5   Reasons for Differences Between United States and Other Countries in Industrial Energy Consumption Relative to GDP, 1972
(tons oil equivalent per $ million GDP)

Countries	Due to structure	Due to energy intensity	Due to inter-action effect	Total actual difference in ratio of industrial energy consumption to GDP
Canada	29.7	−108.7	7.0	−72.0
France	−243.0	265.3	145.5	167.8
W. Germany	−280.1	214.7	135.6	70.2
Italy	−31.7	135.8	10.0	114.1
Netherlands	−64.6	134.3	19.9	89.6
U.K.	−53.1	87.5	10.2	44.6
Sweden	76.8	71.3	−12.4	135.7
Six Western European countries	−153.1	188.2	64.6	99.7
Japan	−138.5	158.0	45.7	65.2

Source: From data in table 6-4.
Note: Appendix E describes the basis for this kind of exercise and the meaning of interaction effect.

explanation—at least in an overall statistical sense—for the high U.S. ratio of industrial energy consumption to aggregate GDP. That is to say, it is a matter of intensities rather than overall structure.

Table 6-5 provides a clearer perspective on this point. The table apportions differences between the U.S. and other-country ratios of industrial energy consumption to GDP in terms of either structure or intensity. The table should be read in the same way as that accompanying the exercise on transport in chapter 5. Thus, taking the six-country Western European grouping as an example, one can ask what accounts for the fact that, per million dollars of GDP, U.S. industrial energy consumption exceeds Western European consumption by 100 tons oil equivalent. The table suggests that it is the fact that not only those 100 tons but an additional 88 tons (188 minus 100) stem from a higher U.S. intensity. This higher U.S. intensity "overexplains" the difference in the ratio. The overexplanation is offset by the fact that the United States consumes less than it would if its industrial GDP were the same proportion of total GDP as Western Europe's. For if the United States were characterized by the Western European structure, it would consume 153 tons oil equivalent more (shown by the minus sign in the first column of table 6-5).

The table indicates that in six of our eight binary comparisons, a similar pattern prevails. That is, the United States would consume less than it actually does at the other country's intensity but more than it does at the other country's proportionate industrial GDP share. In five of these six cases the positive intensity outweighs the negative structural effect.

There are two atypical cases. In the comparison between United States and Sweden, the higher U.S. ratio of industrial energy consumption to GDP arises both because of a higher U.S. intensity and a proportionately more important industrial sector. And in comparison with Canada, the U.S. industrial energy/GDP ratio is actually lower, principally because of a higher industrial energy intensity on the part of Canada.[4]

In spite of the fragile statistical base for this exercise, it seems difficult to dismiss the pervasive and systematic pattern disclosed in table 6-5— that is, the higher U.S. industrial energy intensities as the dominant factor in the contribution of the industrial sector as a whole to overall energy/GDP variability among our countries. What can one make of this finding?

To begin with, inferences regarding the significance of energy intensity at the industry-wide level of aggregation had best be guarded. The industry total conceals a wide and heterogeneous range of activities, particularly within manufacturing—the dominant industrial category. (Manufacturing tends to account for between 75 and 80 percent of industrial GDP in our list of countries.) The finding of higher industrial energy intensity in United States than prevails in almost all the other countries is thus compatible, in an extreme conceptual sense, with two totally divergent possibilities: no matter what the comparative intra-industry mixes, U.S. industrial energy intensity is consistently above that elsewhere; or U.S. intensity is identical to, or even lower than, that of other countries for the same industrial activity but has a high industry-wide intensity by virtue of an orientation toward high-intensity industries. Realistically, of course, neither hypothetical extreme is likely to explain the statistical outcome. No doubt there are elements involving *both* intra-industry intensity variability and mix differences (as well as interactive factors). And in subsequent parts of this chapter, we discuss this issue with respect to several specific industrial activities. The finding there seems to point to higher specific energy intensities as a more pervasive phenomenon than differential intra-industrial structural characteristics.[5]

---

[4] In all these pairs of comparisons the decomposition of the ratio involves also an interaction effect, for reasons described in appendix E.

[5] Note that for any defined intra-industrial component—say steel production— it is always possible to *further* decompose what appears to be a homogeneous output

Findings from Other Studies

Although a generalized interpretation, widely applicable to our observed industry-wide intensity differences, eludes us, it might nonetheless prove instructive to summarize the suggested findings from several other research projects and to speculate on their significance in comparison with our own results.

Doernberg[6] draws attention to the sketchy nature of the evidence on comparative industrial energy intensity for the United States and Sweden. On the basis of the available data, he finds consumption per ton of steel to be about equal in the two countries, while aluminum and cement production is less energy intensive in Sweden. Pulp and paper products are estimated to be somewhat more energy intensive there, although this result is judged to be a function of the respective product mix among pulp, newsprint, Kraft paper, other paper, and boards on the one hand, and the energy intensiveness of each of these products on the other. Also Doernberg's inclusion of Swedish use of noncommercial fuels (excluded in his U.S. data) may also tilt the results slightly toward greater energy intensiveness on Sweden's part.

Schipper and Lichtenberg[7] observe that energy intensities in various manufacturing *industries* are not dramatically different in Sweden and the United States, and this is in line with our own industry-wide finding shown in the second column of table 6-4. This result is heavily influenced by Sweden's manufacturing product mix being weighted to energy-intensive components, for a *product-by-product* examination discloses almost consistently lower Swedish intensities, usually because of reduced process heat requirements.

---

into still additional differentiation of "mix." (For example, we might distinguish between ingots and structural shapes.) Thus, what might, at one level of aggregation, be judged to point exclusively to intercountry intensity differences might, with successive disaggregation, introduce a mix characteristic as well.

The way in which this aspect impinges on intercountry comparisons is illustrated by Schipper and Lichtenberg (Lee Schipper and A. J. Lichtenberg, *Efficient Energy Use and Well Being: The Swedish Example* [Berkeley, Calif., Lawrence Berkeley Laboratory, University of California, 1976], pp. 27–37). They point out that a number of important energy-intensive industries in Sweden (for example, paper, chemicals, petroleum refining, and primary metals) require more energy per unit of corresponding GDP (or value added) than do the same American industries. Yet specific products *within* each of these groups tend, in Sweden, to be manufactured with less energy than in the United States. The resolution comes with examination of the internal structure of these "two-digit" industries: in Sweden relatively more of the GDP is concentrated in the more energy-intensive but less processed, "lower value" end of these industries.

6 A. Doernberg, *Comparative Analysis of Energy Use in Sweden and the United States* (Upton, N.Y., Brookhaven National Laboratory, 1975) p. 15.

7 Schipper and Lichtenberg, *Efficient Energy*, pp. 27–37.

Schipper and Lichtenberg, while not analyzing such disparities in depth, do identify two possible causes: first, the existence of more modern equipment in Swedish industry, though the nature of the investment trends and practices giving rise to this circumstance is not explored; second, there is the proportionately greater use of electricity as an industrial energy source in Sweden. (The lower Swedish energy intensity, as calculated by Schipper and Lichtenberg, prevails in spite of the fact that the calculation assigns the fossil fuel input equivalent of electricity [rather than its inherent thermal value] to specific energy uses.) The greater relative utilization of electricity is, in turn, related to Sweden's historically abundant—and, compared to fossil fuels, cheap—hydroelectric resources. Although, as the price data of appendix F show, industrial electricity in Sweden in 1972 cost somewhat more than in the United States, coal was vastly more expensive; and natural gas, a cheap and widely used American industrial fuel, does not exist in Sweden. Schipper and Lichtenberg state that since "Sweden traditionally has paid a high industrial wage, the saving of energy has come about not by direct substitution of labor for energy, but through the substitution of energy management." [8]

A recent Stanford Research Institute (SRI) study compares energy consumption patterns in the United States and West Germany.[9] The SRI analysis points to distinctly higher U.S. energy intensities for a wide spectrum of categories within the industrial sector. These include: food; paper; chemicals; petroleum refining; stone, clay, glass, and concrete products; primary metals in general; and iron and steel in particular. Prominent within this list are industry groups—particularly primary metals, chemicals, paper, and petroleum refining—whose energy costs as a proportion of total production costs typically exceed the industry-wide share by a large margin. Given German industrial energy prices substantially higher than those in the United States (see appendix F), these intensity differentials are very likely a reflection, at least to some extent, of manufacturers' sensitivity and adaptation to energy-induced cost pressures.

For selected industries, SRI measured intensities both on an energy-per-dollar-of-shipments basis and on an energy-per-physical-quantity basis and found the disparity in intensities to apply in both cases.[10] How-

---

[8] Schipper and Lichtenberg, *Efficient Energy*, p. 35.

[9] Stanford Research Institute, *Comparison of Energy Consumption Between West Germany and the United States* (Menlo Park, Calif., SRI, 1975).

[10] The SRI dollar measure of shipments may move the same way as, but is not equivalent to, the "GDP originating" concept in table 6-4 of our study.

ever, in the case of paper and petroleum products, SRI found the U.S.–
German intensity spread to be much greater on a dollar than on a
tonnage basis—a disparity held by the authors to probably reflect "the
higher prices in West Germany for paper and petroleum products. The
higher energy use per ton of petroleum products in the United States may
be due to greater use of catalytic cracking and other processing to yield
a much higher proportion of gasoline." [11]

If the SRI analysis is on the mark, one would conclude that the higher
industry-wide U.S. energy intensities reflect a fairly pervasive intra-
industry phenomenon rather than primarily a product mix characteristic.

A recently published study by Gordian Associates[12] compares, in exhaustive detail, energy consumption in the United States, West Germany,
Italy, and the United Kingdom in the manufacture of a single, relatively
homogeneous product, cement. The data appear below.[13]

	U.S.	W. Germany	Italy	U.K.
Kiln fuels consumed in making cement (overall, based on clinker) (kcal per kg clinker).	1,568	977	1,058	1,379
Electricity consumed in making cement (overall, based on clinker) (kWh per ton clinker).	152	118	123	109

According to these data the United States and the United Kingdom
are reasonably close, both substantially above the energy intensities of
West Germany. Part of these differences are due to differences in size
and age of plant. In addition, the type of process used affects energy
intensities. This same report estimates on the basis of "best practice/large
plants" that wet processes consume about 50 percent more energy per
ton of cement than dry processes, although other sources give a considerably reduced differential. The United States and the United Kingdom
both have a much higher wet capacity (59 and 60 percent, respectively)
than Germany and Italy (5 and 13 percent). These differences in capac-

---

[11] Stanford Research Institute, *Comparison of Energy*, p. 52. Ideally, of course,
an intensity comparison using values in the denominator should correct for, rather
than reflect, the price differentials mentioned in the quoted SRI passage. Such correction is the virtue of the purchasing-power-parity concept as discussed in chapter 2
of our study.

[12] Gordian Associates, Inc., "Industrial International Data Base Pilot Study: The
Cement Industry" (New York, Committee on the Challenges of Modern Society of
the North Atlantic Council, 1976).

[13] Data refer to the production of Portland Cement and do not make allowances
for reductions in energy consumption per ton of product by the use of materials to
make blended cement.

ity would therefore explain many of the differences in energy intensities between these pairs of countries. But it is important to note that the process chosen is influenced by other raw material characteristics and economic factors. In the words of the Gordian report, ". . . it is clear that the higher energy consumption shown by the wet process is a less important item when energy costs are low, as used to be the case in the United States. The historically high energy costs in Europe and Japan have encouraged adoption of pre-heater processes, with their attendant higher energy efficiency." [14] In other words, although the higher energy intensity of the U.S. cement industry appears to be well established, the choice of process, at least in 1972, was consistent with historical patterns of energy price differentials.

Price and unit value indexes for fuels and power purchased by industry are given in table 6-6. This indicates that for industrial use, energy prices in most countries are generally about double the U.S. levels. The exceptions are Canada—only 50 percent higher—and, somewhat ambiguously, Sweden, which might be nearer the Canadian level or the other-country level, depending on the weighting system used. Note that our price data apply in principle to prices paid by the industrial consumers, including taxes. In Europe in 1972, taxes ranged from 15 to 30 percent on heavy oils, compared with zero excise taxes in the United States.

This lower U.S. price fits well with the relatively high energy intensity of U.S. industry, but the relationship is less well established for the other countries in which energy intensities vary much more widely than do the price differences between them.

Data contained in a recent UN document, while throwing light on the energy intensities characterizing various industry sectors, also serve to underscore the difficulties of interpreting the significance of intercountry intensity differences.[15] The figures shown in table 6-7, said by the UN to apply to "around" 1965, are ratios of the value of energy inputs of an industrial sector to the value of output of that sector, where the values for a given country reflect its own national currency. In other words, the figures are direct input–output coefficients. We have chosen to show only those that are typically among the high-coefficient industries.

The figures are useful in alerting us to the identity of those industries and their relative energy-intensity rank within the country in question. As we would expect, the energy industries themselves lead the pack by a wide

---

[14] Gordian Associates, "The Cement Industry," p. 30.

[15] U.N. Economic Commission for Europe, *Increased Energy Economy and Efficiency in the ECE Region* (New York, UN, 1976).

TABLE 6-6 Energy Prices in Industry, 1972
(absolute figures in dollars per million kcal)

Prices	U.S.	Canada	France	W. Germany	Italy	Netherlands	U.K.	Sweden	Japan
Fuel									
Electricity	11.0	17.7	21.9	28.7	27.2	26.1	24.9	15.0	25.4
Gas	1.7	1.7	3.1	3.3	3.0	2.1	2.3	a	a
Coal	1.8	1.9	3.9	3.9	4.7	3.4	3.9	3.6	3.6
Petroleum products	2.8	2.2	3.7	2.9	3.5	2.5	4.3	2.7	3.2
Industrial sector									
pure price index (U.S. = 100)	100.0	131.5	182.6	210.3	204.8	188.3	193.0	138.7	203.1
unit value index (U.S. = 100)	100.0	158.4	185.6	202.8	206.1	167.8	183.5	187.2	225.1

a Negligible consumption.
Source: Appendix F and **RFF** estimates.

TABLE 6-7   Direct Energy Coefficients: Ratio of the Value of Energy Inputs to the Value of Output of that Sector, Selected Countries, Around 1965

(units of energy per 100 units of output based on national monetary units)

	U.S.	France	W. Germany	Italy	Nether- lands	U.K.	Sweden
Energy industries	35.9	24.9	32.3	26.6	40.2	34.2	4.7
Basic metals	4.3	8.7	8.6	7.9	5.5	14.5	6.1
Chemicals	5.19	7.4	6.9	8.8	9.0	9.0	4.6
Non-energy mining	5.20	5.1	16.9	7.8	—	10.8	3.9
Wood/paper and products	1.6	3.1	3.1	3.2	2.2	2.1	3.1
Construction	2.2	2.0	1.0	1.6	3.0	0.9	1.3
			Ranking				
Energy industries	1	1	1	1	1	1	2
Basic metals	4	2	3	3	3	2	1
Chemicals	3	3	4	2	2	4	3
Non-energy mining	2	4	2	4	—	3	4
Wood/paper and products	6	5	5	5	5	5	5
Construction	5	6	6	6	4	6	6

Source: UN Economic Commission for Europe, *Increased Energy Economy and Efficiency in the ECE Region* (New York, UN, 1976) pp. 14–15.

margin. (Sweden is low because of the absence of energy extraction and a rather small refining sector.) Typically, basic metals, chemicals, and non-energy mining follow high on our list of six industry groups, while the wood and paper and construction components tend to trail somewhat.

But perhaps the most instructive thing to observe in table 6-6 is the extent to which these coefficients—being based on value relationships reflecting national currencies—conflict with the observations earlier in the chapter. For although our previous discussion seemed to point to inherently high comparative U.S. energy intensities, table 6-7 discloses at best comparable, and in numerous instances, lesser intensities. Most likely, it is the absence in the table of an adjustment reflecting international disparities in energy and nonenergy prices that produces this apparent anomaly. Since foreign prices for energy tend to exceed U.S. prices by more than foreign prices for other things, an adjustment for that situation would lower the foreign coefficients below those shown in table 6-7—probably causing them no longer to conflict with the earlier findings regarding higher U.S. industrial energy intensities.

Energy and Labor Productivity

Before proceeding to a number of discrete industrial-energy topics, we might close out this section of the chapter by emphasizing, as we have done elsewhere in this study, the restricted significance of comparative energy intensities. This caveat applies more forcefully to the industrial sector than to any other sector in our study. Energy represents only one component of industrial costs. We have ignored such complementary inputs as labor, capital, and other natural resources. Few technologies or production processes are so inflexible that they do not permit some "input juggling" depending on relative costs of the different factors. In other words, differences in energy intensities—especially minor differences—need not betray differences in *overall* efficiency. The whole question of how energy intensity relates to economic efficiency has not yet received much attention, particularly of an empirical sort.

The one aspect of this issue which we can illuminate slightly has to do with the triangular interconnection among labor productivity, the amount of energy associated with each worker, and energy/GDP ratios. It is an unarguable proposition that in a sophisticated industrial economy, production efficiency depends on some important—if quantitatively unspecified—interaction of labor and energy-using machines or processes. The interesting thing is to find a much narrower range of variability in industrial output per worker than in industrial energy consumption per worker—hence, the numerical result of high North American energy intensities (see table 6-8). Clearly, Europe and Japan's recorded labor

TABLE 6-8   Energy Consumption and Labor Productivity, 1972

GDP and energy consumption	U.S.	Canada	Five W. European countries	Japan
GDP per employee (dollars)	14,899	12,079	7,862	8,135
Index (U.S. = 100)	100.0	81.1	52.8	54.6
Energy per employee (tons oil equiv.)	21.3	21.5	7.2	7.5
Index (U.S. = 100)	100.0	100.9	33.8	35.2
Energy per unit of industrial GDP (tons oil equiv. per $ million GDP)	1,427.7	1,777.2	918.6	924.6
Index (U.S. = 100)	100.0	124.5	64.3	64.8

Note: France is excluded from the Western European grouping because of suspect data.

Source: Employment data from Organisation for Economic Co-operation and Development, *Labor Force Statistics, 1961–1972* (Paris, OECD, 1974). Energy and GDP figures are based on data in appendixes B and A, respectively.

productivity compared with that of the United States does not seem to be dependent on commensurate amounts of energy available to the industrial work force. Up to a point, and in comparison with other countries, there may be some degree of substitution of U.S. energy for U.S. labor in order to produce a given level of output. However, a very important observation is made in the SRI study.[16] There we are reminded that the major portion of U.S. industrial energy use is for process heat and space heating, rather than power. In some fashion, other factors are at work—so to speak—in bringing about the results pictured in table 6-8.

Industry Profiles

We have selected certain industries for particular attention: iron and steel, chemicals, and petroleum refining. They have been chosen on the basis of data availability and at least reasonable homogeneity of output. Although they may not be representative of the industrial sector as a whole, they are all very energy-intensive industries and major industrial users.

Iron and Steel Industry

The iron and steel industry is invariably the largest single industrial consumer of energy in our countries; the industry normally accounts for between 15 and 25 percent of all energy consumed in the industrial sector (including the energy industry), and between 5 and 10 percent of total energy consumption.[17]

In addition to being a major consumer, the iron and steel industry produces a reasonably homogeneous output. This is an advantage for analytical purposes because it means that total differences in energy consumption relative to GDP among our countries can be explained by relative size of industry, and by the energy intensity of the product (measured by energy input per ton of crude steel). In other words, it is assumed that there is no variation in product mix within steel which might contribute to differences in energy consumption in iron and steel relative to GDP. As will be seen later, this is a great oversimplification; nonetheless, compared with the wide range of product mix of other industries—for example, the chemical industry, another major energy consumer—the output of the iron and steel industry is relatively homogeneous.

---

[16] Stanford Research Institute, *Comparison of Energy*, p. 52.
[17] "Total inland consumption," as defined by OECD.

ACCOUNTING PROCEDURES. Before proceeding with our statistical analysis it is necessary to define some terms. Differences in results that appear in the several studies made of energy consumption in steelmaking have been frequently found to stem from differing definitions of energy input and product output. It is therefore essential in this study to be clear about what has been included and, perhaps more important, what has not been included.

First, our energy consumption data apply to the whole of the iron and steel industry, including the large amounts of energy used in rolling and finishing steel, whereas our measure of output is crude steel rather than output of finished products.

Second, energy consumption is confined—as it is throughout this report—to direct consumption of energy, except to the extent that energy to make pig iron is itself an exception to this practice. This includes, for example, the coal and gas (including blast furnace and coke oven gas), petroleum products, and electricity used in iron and steelmaking. It *does not include* energy used up in the cokemaking process, energy contained in scrap, or waste heat incurred in electricity generation. Following our system, such losses have already been accounted for in the consumption of the energy sector and transformation losses, respectively, or, in the case of the energy content of scrap, not at all. If waste heat had been included, the iron and steel industry energy consumption data for those countries which are heavy users of electricity in their iron and steel industries (Italy, Sweden, and Japan) would have been higher, relative to other countries. This effect could, however, have been offset by a high proportion of self-generation, which permits, though not always ensures, some recuperation of waste heat.

VARIATIONS IN ENERGY CONSUMED RELATIVE TO GDP. Based on these definitions, energy consumption by the iron and steel industry relative to GDP is given in table 6-9. In most of our countries, including the United States, consumption of energy in iron and steelmaking relative to GDP is very similar. The three exceptions are Germany, United Kingdom, and Japan which consume substantially more. Since we assume for the moment that the products of steelmaking are relatively homogeneous, the difference between countries in energy consumption relative to GDP can be explained by differing size of industry or varying energy intensity in the production process.

As can be seen in table 6-9, the relative size of the industry (measured by tons of crude steel output per million dollars of GDP) varies consid-

TABLE 6-9 *Energy Consumption in the Iron and Steel Industry and Related Data, 1972*

Item	U.S.	Canada	France	W. Germany	Italy	Netherlands	U.K.	Sweden	Japan
Energy consumption iron & steel industry (million tons oil equiv.)	68.87	4.63	12.86	24.38	7.95	2.76	15.55	2.42	45.87
As percent of total industrial energy	13.2	8.7	21.6	26.5	16.9	15.9	20.5	19.3	33.4
As percent of total energy consumption	3.9	2.5	7.5	9.6	6.1	4.4	7.3	5.6	14.9
Production of crude steel (million tons)	120.88	11.87	24.05	43.71	19.81	5.60	25.32	5.26	96.9
Steelmaking capacity									
Percent open-hearth [a]	26.2	39.5	43.5	25.2	20.1	4.2	37.9	21.3	2.0
Oxygen	56.0	43.9	45.9	64.6	39.1	88.9	42.7	36.5	79.4
Electric arc	17.8	16.6	10.6	10.2	40.8	6.9	19.4	42.2	18.6
Energy consumption of iron & steel relative to GDP (tons oil equiv. per $ million GDP)	58.4	44.8	59.7	100.9	55.9	56.3	81.8	59.5	126.5
Crude steel production (tons per $ million GDP)	103	115	112	178	139	114	133	129	267
Energy consumption (tons oil equiv. per ton crude steel)	0.57	0.39	0.53	0.56	0.40	0.49	0.61	0.46	0.47

Source: Production data from Organisation for Economic Co-operation and Development, *The Iron and Steel Industry in 1972 and Trends in 1973* (Paris, OECD, 1974); other data from appendix B.
[a] Includes basic and acid Bessemer.

erably from country to country. The United States has the smallest, and Japan (two and a half times the United States), by far the largest. Relative size of industry is therefore responsible for some of the differences in iron and steel energy consumption relative to GDP, particularly the high Japanese consumption, but it does not explain the relative position of the United States. If it did, the United States would have the lowest energy consumption in iron and steel relative to GDP.

Variation in energy consumption per ton of output must also contribute. The United States, the United Kingdom, and West Germany appear to use relatively large quantities of energy per ton of steel produced, and Japan, Netherlands, and Sweden, relatively small quantities. These results appear to be in accord with the results of other studies[18] except for Germany.

Most of these studies cover the United States and one other country, although one, prepared by the Battelle Columbus Laboratories, covers four. All use different accounting systems and so are not directly comparable either with each other or with our data. For that reason, the energy consumed per ton of crude steel is given below in indexes of U.S.=100 to illustrate the relative difference between the United States and the other countries.

Study	U.S.	W. Germany	U.K.	Sweden	Japan
RFF	100	99	100	81	82
Battelle	100	87	100	19	76
SRI	100	74	19	19	19
Schipper	100	19	19	69	19

The average-to-low position of the United States, in terms of energy consumption in iron and steel relative to GDP, comes about through a combination of two opposing elements—the smallest iron and steel industry relative to GDP, and one of the highest energy intensities per ton of steel produced. This is set out more systematically in table 6-10, where the total difference between countries in energy consumption in the iron and steel industry relative to GDP is partitioned into the effect caused by size of industry and that caused by differing energy intensity. The

[18] Battelle Columbus Laboratories, *Potential for Energy Conservation in the Steel Industry* (Columbus, Ohio, Battelle Columbus Laboratories, 1975); J. A. Over, ed., *Energy Conservation: Ways and Means* (The Hague, Future Shape of the Technology Foundation, 1974; P. F. Chapman, "The Energy Cost of Materials," *Energy Policy* vol. 3, no. 1 (March 1975); Stanford Research Institute, *Comparison of Energy;* D. J. Wright, "Natural Resource Requirements of Commodities," *Applied Economics* vol. 7 (1975) 31–39; Lee Schipper and A. J. Lichtenberg, *Efficient Energy.*

[19] Not available.

TABLE 6-10 Reasons for Differences Between United States and Other Countries in Energy Consumed in the Iron and Steel Industry, Relative to GDP, 1972
(tons oil equivalent per $ million GDP)

Differences	Canada	France	W. Germany	Italy	Netherlands	U.K.	Sweden	Japan
Total difference from U.S.	13.6	−1.3	−42.52	2.49	2.1	−23.42	−1.09	−68.10
Difference due to structure	−5.89	−5.28	−43.15	−17.77	−6.20	−18.20	−13.62	−85.94
Difference due to energy intensity	19.49	3.98	0.63	20.26	8.31	−5.22	12.53	17.84

Note: The interaction effect has been apportioned roughly between differences due to structure of the industry, and differences due to energy intensity.
Source: Estimated on the basis of data in table 6-9.

higher U.S. energy intensity—measured in energy consumption per ton of crude steel—accounts for all, and more, of the total difference between the United States and other countries. The sole exception is the United Kingdom whose higher energy consumption in iron and steel relative to GDP compared with the United States is due both to a larger iron and steel sector and also to higher energy consumption per ton of steel.

SOURCES OF VARIATIONS IN ENERGY INTENSITIES IN IRON AND STEEL-MAKING. Variations in energy consumption can occur in many stages of steelmaking. The operation of blast furnaces, which produce pig iron from iron ore, coke, sinter, and limestone, accounts typically for some 50 percent of the total energy consumed in the steelmaking process. But the amounts of energy needed in blast furnace operation can vary by as much as 10 percent, depending on the quality of iron ore used. In general, the higher the quality used, the less energy needed. Canada and Sweden are the only countries that produce high-grade iron ore. Other countries can and do import ore of high quality, but if countries also have domestic deposits of low- and medium-grade ore as in the case of the United States and France, these tend also to be used. Those countries with the lowest energy intensities—Italy, Netherlands, and Japan—import virtually all the ore they require, although it is not possible to confirm from import data the quality of the ore imported. A second source of differing energy consumption in the iron-making process stems from the proportion of pig iron and scrap requirements that can be obtained domestically. As the blast furnace accounts for such a high percentage of the total energy consumption, significant savings in energy consumption can be achieved by importing pig iron or scrap rather than importing iron ore. Most of our countries meet virtually all of their pig iron requirements from domestic production. Italy and Sweden, however, import 11 to 13 percent of their pig iron requirements, thus avoiding part of blast furnace energy consumption. With regard to scrap, Italy imports 50 percent of its scrap requirements, and Canada more than 10 percent.

These two factors—quality of ore used and proportion of pig iron imports relative to consumption—affect only those countries with lower intensity than the United States. Together they may account for roughly one-half of the difference in energy consumption per ton of steel relative to the United States.

A further source of variation in steel industry energy intensities is the type of furnace used in making steel. The process at this stage accounts for 15 to 20 percent of total energy used in iron and steel manufacture.

The three main types of steelmaking furnaces are: open-hearth (including basic and acid reserve), oxygen, and electric arc. Each process involves different energy usage so that the amount of energy various countries use for making a ton of steel can be affected by the mix of different processes.

Variations in the amount of energy consumed per ton of steel using the different processes also depend on the energy accounting procedure used. Following our system, the energy consumed in each process includes energy at the heat delivered rate (that is, heat losses are excluded), excludes the energy content of scrap, excludes the energy embodied in the blast furnace metal, excludes energy used in oxygen making, and excludes steam produced in the steelmaking process (which may or may not be captured depending on the heat efficiency of the system). Following this procedure, the open-hearth furnace is the most energy intensive—over twice as energy intensive as the electric arc and the basic oxygen. It should again be emphasized that this approach penalizes the open hearth; if heat losses had been included, the electric arc would have been the most energy intensive. In addition, the inclusion of the energy content of scrap would have further added to the overall energy intensiveness of the electric arc.

Given these definitions, those countries which have a larger proportion of steelmaking capacity in open-hearth furnaces relative to the United States will therefore tend to have a higher consumption of energy per ton of iron and steel produced. Among our countries the United States has a relatively large proportion of open-hearth capacity exceeded only by Canada, France, and United Kingdom. This factor then would seem to account for the higher U.K. and U.S. consumption per ton, but seems inconsistent with the very low Canadian consumption. On the other hand, those countries with negligible open-hearth capacity—Netherlands and Japan—have relatively low energy consumption per ton of crude steel.

A final stage in iron and steelmaking is the rolling, reheating, and shaping of steel. This accounts for about 25 percent of total energy used in making steel, varying with the degree of continuous casting and the final mix of semifinished and finished products.

Continuous casting can save about 20 percent of the total energy in this final process. The proportion of continuous casting in the total varies considerably between countries. In the United States and the United Kingdom, for example, only 6 and 3 percent respectively of crude steel is continuously cast, compared with 17 and 21 percent in West Germany and Japan. This type of variance could make a difference of some 3 percent in total energy needs of the iron and steel industry in the United Kingdom and United States, compared with West Germany and Japan.

Finally, there is a considerable variation in energy needs of the different finished products. Some of the more energy-intensive finished products use five or six times more energy per ton than the less energy-intensive products. Because of the wide variety manufactured, it was not possible to assess the net effect of different finished product mix on energy consumption per ton. But from data available it appears that whatever the difference in product mix, it is unlikely to account for much of the relatively high U.S. specific consumption.

In conclusion: U.S. consumption of energy in iron and steel relative to GDP stands at about the average for all our countries. But the United States has a small iron and steel sector relative to the other countries and a higher consumption of energy per ton of crude steel produced. This high consumption per ton comes about through a relatively large proportion of the high energy-using, open-hearth process and a relatively small proportion of continuous rolling capacity. Both features are shared by the United Kingdom, which has similarly high energy consumption per ton of steel. Italy, Netherlands, Sweden, and Japan, the countries with low energy consumption per ton, have certain characteristics in their iron and steel industries not shared by the high energy users. Italy, for example, imports substantial quantities of pig iron and scrap, therefore avoiding part of the energy-expensive blast furnace process; all are assumed to use higher quality ores; the Netherlands and Japan have virtually none of the energy-intensive, open-hearth steel furnaces; Italy and Sweden have an exceptionally high proportion of electric arc furnaces, the least energy intensive of all following our accounting procedures; and Japan is known to have a relatively high proportion of continuous casting. These factors together would in most cases account for much of the difference between United States and other-country energy consumption, although Italy would still remain significantly lower.

Reference has already been made at various points to the possibility that energy consumption associated with the iron and steel industry is affected by the differing amounts of energy embodied in trade. As previously mentioned, this would mean that importing high quality rather than low quality ores could result in reduced energy needs, and the importing of pig iron could bypass the highly energy-expensive blast furnace stage.

At the stage of the finished product, too, trade can explain difference in industry size and therefore difference in energy consumption relative to GDP. All countries have substantial trade in both semifinished and finished products, but in most cases imports are fairly closely matched by exports so that the net balance is small in relation to total consumption. The United States is the sole country with substantial net imports (equal to 8 percent of total consumption), but three countries—United Kingdom, Netherlands, and Japan—have substantial net exports accounting for 9, 36, and 27 percent respectively of total consumption. Thus the smaller size of the U.S. industry noted earlier and the relatively large size of the U.K. and Japanese industries are closely associated with the proportion of supplies imported or exported.

The Chemical Industry

Energy is consumed by the chemical industry both in the form of fuel and power and also as industrial feedstock.[20] Together, chemical industry use for both purposes probably accounts for 6 to 8 percent of total energy consumption, with the exception of the Netherlands and Italy, whose chemical industries utilize a much larger part—about 20 percent of total energy consumption (see table 6-11). (Note that previously nonenergy uses have not been included in industrial energy consumption. They are included here because they comingle with fuel and power use in chemicals.)

The United States appears to consume significantly more energy (both as energy and as feedstock) in its chemical industry relative to GDP than the other countries, with the exception of the Netherlands and Italy, who consume very much more.

The question is how much of the higher consumption of these countries is due to larger industries or, on an aggregative basis, to more energy-intensive processes. The latter question, in turn, hinges on the mix within the chemical industry of energy-intensive products. It is, unfortunately, not possible to analyze these differences in the chemical industry in the same way as iron and steel and petroleum refining because of the difficulty of obtaining a measure of output for the chemical industry. This industry

[20] The consumption of feedstocks in the chemical industry (as compared with other industries) was estimated on the basis of the composition of total nonenergy uses, using data from the European Communities, *Energy Statistics Yearbook 1969–73* (Luxembourg, European Communities Statistical Office, 1974). The remainder consists of such items as bitumen, road oil, petroleum, coke, lubricating oils, and white spirit.

TABLE 6-11 *Energy Consumption in the Chemical Industry and Related Data, 1972*

Item	U.S.	Canada	France	W. Germany	Italy	Netherlands	U.K.	Sweden	Japan
Energy consumption in chemical industry (million tons oil equiv.)	70.00	0.98	6.24	12.44	12.00	6.28	5.82	0.57	12.33
Feedstock consumption in chemical industry (million tons oil equiv.)	56.56	3.98	4.27	5.98	9.79	8.35	6.71	0.50	13.07
Total energy consumption in chemical industry including feedstock (million tons oil equiv.)	126.56	4.96	10.51	18.42	21.79	14.63	12.53	1.07	25.40
Percentage of total energy consumption	7.2	2.7	6.1	7.3	16.7	23.5	5.8	2.4	8.2
Total energy consumption including feedstock (tons oil equiv. per $ million GDP)	107	47	49	75	153	298	66	26	70
Difference from U.S. (tons oil equiv. per $ million GDP)	—	63	58	32	−46	−191	41	81	37
Value of shipments (billion U.S. dollars)	51.95	n.a.	9.52	14.30	7.96	2.99	14.49	1.03	19.95
Value of shipments as percent of GDP	4.4	n.a.	4.4	5.8	5.6	6.1	6.7	2.5	5.5
Energy consumption including feedstock (tons oil equiv. per $ million of shipments)	2436	n.a.	1105	1288	2737	4890	864	1040	1271

Note: Blank = not applicable.
Source: Value of shipments from Organisation for Economic Co-operation and Development, *The Chemical Industry, 1972–73* (Paris, OECD, 1974); other data from appendix B.

produces such a wide range of products—industrial chemicals, man-made fibers, fertilizers, dyestuffs, paints, plastics, soaps and detergents, and pharmaceutical and cosmetic preparations—that it is impossible to arrive at a volume measure of output.

For some countries, a value measure of output, "value of shipments," was available. These data were converted into U.S. dollars using the binary ideal rate in an effort to reduce the distortions of real output measurement inherent in using market exchange rates (though we do not know at this level of disaggregation whether this purpose is achieved). This still leaves the problem of price differences among countries—a problem that becomes potentially more troublesome, the smaller the sector under consideration. The data on value of shipments used here, therefore, are only an imperfect indicator of comparative real output of the chemical industries among our countries.

According to these data, the U.S. chemical industry is one of the smallest, relatively speaking, of the countries considered here. That is to say, the large consumption of energy, including energy for feedstock by the U.S. chemical industry relative to GDP, is *not* due to the larger size of the industry but rather to a higher energy consumption per unit of aggregate chemicals produced compared with other countries. These results are borne out, at least so far as the United States is concerned, by other studies comparing the United States with West Germany[21] and Sweden.[22]

Part, though not all, of the higher Italian and Netherlands consumption is explained by a larger industry, but energy consumption per unit of output still remains substantially above other countries and even above the United States. The relatively high energy intensity of all three countries is related to the concentration of natural-gas-based chemical industries associated with domestic production of natural gas.

The difference in chemical industry energy consumption relative to shipments could be due either to a more energy-intensive intrachemical product mix or to an inherently more energy-intensive technology, or to both. It is beyond the scope of this report to make a systematic examination of these elements. A recent, very detailed and highly disaggregated analysis of polymers discloses no systematic variability in intensities. For example, the United States appears to be over twice as energy intensive as the Netherlands in production of polyethylene; roughly comparable in intensity in production of PVC (polyvinyl chloride); and

[21] Stanford Research Institute, *Comparison of Energy.*
[22] Schipper and Lichtenberg, *Efficient Energy.*

moderately less energy intensive in polystyrene. Differences in basic technology, in process-heat applications, and in innumerable other complex production characteristics (some of which differentiate plants producing the same product *within* a country as well as plants between countries) are discussed in the Berry study, but no generalized conclusion emerges.[23]

Just as in steelmaking and petroleum refining, the size of the chemical industry depends not only on the level of consumption but also on imports and exports. The Netherlands and West Germany both have net exports equivalent to 30 and 20 percent of production respectively, whereas Sweden has net imports of more than 33 percent. This aspect enters into the discussion of energy-intensive foreign trade commodities later in this chapter.

Petroleum Refining

Among the energy industries, petroleum refining is the single largest[24] user of fuel and power, accounting in most countries for between 5 and 10 percent of total industrial use (including the energy industry).[25] The Netherlands consumes a substantially higher proportion—16 percent— and Canada and Sweden much lower, 2 percent (see table 6-12).

Relative to GDP, energy consumption by petroleum refineries varies widely. The Netherlands is by far the largest consumer relative to income followed by the United States. Well below the United States, the other countries are fairly closely grouped, with the exception of Sweden, whose consumption relative to GDP is one-tenth that of the Netherlands.

Following our usual procedure, these variations in energy consumed in petroleum refining relative to income can be attributed either to size of the industry or to the energy intensity of the operation. As table 6-12 shows, the United States has one of the smallest refinery industries, as measured by output relative to GDP. Only Germany and Sweden are smaller, while the Netherlands is three times larger. The size of a country's petroleum refining capacity is closely related to its trade in petroleum products. If countries import products rather than refine at home, fewer

---

[23] R. Stephen Berry, Thomas V. Long III, and Hiro Makino, "An International Comparison of Polymers and Their Alternatives," *Energy Policy* vol. 3, no. 2 (June 1975).

[24] For those countries with domestic natural gas production—the United States, Canada, and Netherlands—gas consumption in natural gas extraction is also a major end use.

[25] These figures do not include consumption of by-product refinery gases, which typically account for almost one-half of the total.

TABLE 6-12 Energy Consumption in Petroleum Refining and Related Data, 1972

Item	U.S.	Canada	France	W. Germany	Italy	Netherlands	U.K.	Sweden	Japan
Energy consumption in petroleum refineries (million tons oil equiv.) [a]	62.09	2.51	7.49	8.52	3.87	3.45	7.08	0.32	9.69
As percent of total consumption in energy and industrial sectors	9.0	3.3	9.9	7.1	6.3	15.7	7.3	1.6	5.0
Energy consumption in refineries (tons oil equiv. per $ million GDP)	52.7	24.3	34.8	34.6	27.2	70.4	37.3	7.9	26.7
Difference from U.S.	—	18.4	17.9	18.1	25.5	−17.7	15.4	44.8	26.0
Total refinery output (million tons)	541.8	73.5	110.5	102.8	110.1	66.1	99.4	10.9	186.0
Energy consumed as percent of refinery output	11.5	3.4	6.8	8.3	3.3	5.2	7.1	2.9	5.2
Refinery output (tons oil equiv. per $ million GDP)	460	711	513	418	817	1348	523	268	513
Net imports of petroleum products	128.3	3.3	(3.1)	26.9	(22.6)	(28.1)	2.9	17.2	17.9
Consumption of petroleum products	668.9	77.3	90.2	119.1	69.4	21.0	78.8	24.1	154.0
Net imports as percent of consumption	18.5	4.3	(3.5)	22.6	(32.5)	(133.98)	3.7	71.4	11.6
Gasoline as percent of total output	44.6	28.1	12.5	15.2	11.9	8.0	13.7	11.6	10.4

Note: Figures in parentheses are net exports. *Blank* = not applicable.

Sources: Data in appendix B; Organisation for Economic Co-operation and Development, *Statistics of Energy 1959–72* (Paris, OECD, 1974).

[a] Excludes consumption of refinery gases which might account for one-half of the total.

refineries will be needed. Similarly, countries with a large export trade in products will have a larger petroleum refining industry than those which refine only for domestic needs. Thus, the relatively small size of the U.S. and German refining industry is partially due to the fact that net imports of products represent 20 percent of petroleum consumption; and the very small size of the Swedish sector is due to that country's importing 70 percent of its petroleum products. The large Netherlands sector, on the other hand, is explained by the large net export trade in refined products, 130 percent higher than domestic consumption.

As the United States' large refinery consumption relative to GDP cannot be explained by a relatively large petroleum refining industry, energy intensities (measured by energy consumed in refining per unit of refinery output) must be responsible. As table 6-12 shows, the United States is by far the most energy intensive, consuming an average of about double the energy per unit of output of the European countries. Deficiencies in the data may explain part of this difference.

Difference in energy intensities may also be due to the contrast in product mix between countries, which is obscured by our level of aggregation. The United States is unique, even compared with Canada (whom it resembles in so many aspects of the energy economy), in that it has an exceptionally high proportion of motor gasoline in the total output—40 percent compared with an average of 10 percent in European and Japanese refineries. The higher proportion of product mix in highly refined products requiring catalytic cracking is responsible for much of the higher energy intensity of petroleum refining in the United States.

Special Topics

In this chapter, and others, attention is constantly drawn to the deficiencies of energy consumption data. In addition to the usual difficulties of comparability, there are certain aspects of energy consumption which, although not strictly speaking definitional in nature, do affect the amount of energy consumed and which could, therefore, play some role in explaining differences in energy consumption between countries. Here we examine two such factors—energy embodied in nonenergy foreign trade, and the effect of fuel mix on industrial consumption.

Energy Embodied in Nonenergy Trade

In this chapter, variations in industrial energy consumption relative to income have been partitioned into structural and intensity effects, in which "structure" is measured by the size of the sector relative to total output and "intensity" by the amount of energy used in producing a unit of output. It will be recalled that, in comparison with the United States, most countries had a relatively larger industrial sector.

But this larger industrial sector is in many cases associated not so much with higher domestic consumption levels as with a substantial export trade in manufactured products. That is to say, some, perhaps an important part, of the energy consumed in industry is shipped outside the country embodied in exports of manufactured and other goods. Analogously, part of a country's effective consumption of energy comes from the energy embodied in imports. Because of this, our total energy consumption may give a misleading view of the amount of energy effectively consumed within a country. (Note that we are talking about the energy content of *nonenergy* trade: the energy content of energy trade has already been taken into account in total energy consumption, which includes the energy content of fuel imports and excludes the energy content of fuel exports.)

It is of interest, therefore, to estimate the energy content of trade in nonenergy goods in relation to each country's total energy consumption; and to see whether these indirect flows are significant enough to explain the disparity in consumption levels between the United States and other countries. The potential existence of considerable differences in the net flows of embodied energy is indicated by the differences in size of trade sectors among our countries. The United States and Japan, for example, have relatively small trade sectors: imports account for 6 to 8 percent of GDP compared with the more typical European level of 20 percent. Netherlands is exceptionally high at over 40 percent. In order to arrive at a closer approximation of trade in nonenergy items, we have excluded trade in fuels. On this basis, three countries, the United States, Canada, and the United Kingdom turn out to have small import surpluses and the rest, export surpluses, rising to about 4 percent of GDP in the case of Sweden, Germany, and Japan. If the energy content of nonenergy trade bears any relationship to the value of nonenergy trade, we should therefore expect United States, Canada, and United Kingdom to be small net importers of energy embodied in nonenergy foreign trade and the other countries, net exporters of varying degree.

However, the relationship between the value and energy content of nonenergy trade may not be as simple as that. The energy intensities of different classes of imports and exports can vary considerably. Depending on the structure of each country's trade, which again varies widely among countries, both the magnitude and direction of energy flows could be changed. In order to have a more reliable estimate of the energy content of nonenergy foreign trade, it is necessary to disaggregate the trade data and apply to each category an appropriate energy intensity.

Although energy intensities for imports and exports no doubt vary from country to country, it was not possible to obtain the data for each of our countries. Instead, energy intensities for the United States were applied to the trade of all countries, which implicitly assumes that all countries have the same technology as the United States. Insofar as the energy intensities of other countries might be lower than those of the United States, then the energy content of their foreign trade would be less than estimated here.[26]

The energy intensities were multiplied by the value of trade in the corresponding items or categories to give the energy content of each country's nonenergy imports and exports. The results appear in table 6-13. Of all our countries only the United States and France import more embodied energy than they export, but in both cases the net amount is very small, equivalent to about 1 percent of total energy consumption. All other countries export more embodied energy than they import, but again in several cases, Canada, Netherlands, and the United Kingdom, the amounts are very small—about 1 percent of total energy consumption.

The remaining countries have more substantial net exports, Italy about 5 percent and Sweden, West Germany, and Japan as much as 12 percent. To revert to the original question of how far the different amounts of energy embodied in nonenergy foreign trade might explain the higher U.S. energy consumption, the answer must be that it does not. Quite the reverse: the effect of taking into account energy embodied in nonenergy foreign trade widens the disparity between energy consumption levels in the United States and the other countries.

[26] The energy intensities used are estimates based on data contained in Robert A. Herendeen and Clark W. Bullard III, *Energy Costs of Goods and Services, 1963 and 1967* (Urbana, Ill., Center for Advanced Computation, University of Illinois at Urbana-Champaign, November 1974). This gives the energy content per dollar of final output of a wide range of goods and services for finished products. The range of energy intensities between food and raw materials and manufactured goods is not as large as might have been anticipated, presumably because of the higher value added of manufactured goods. Even so, there is some tendency to higher intensities per unit of output in manufactured goods. We have assumed 15,000 kcal per dollar for raw materials and 25,000 for manufactured goods. Within these two broad groups, some items or groups of items stand out as being more energy intensive than the group generally—chemicals, paper and board, textiles, and iron and steel. These items have been assigned individual energy intensities. (Chemicals 40,000 kcal per dollar; textiles 30,000; iron and steel 45,000; and nonferrous metals 40,000.)

TABLE 6-13 *Energy Embodied in Nonenergy Foreign Trade, 1972*
(million tons oil equivalent)

	U.S.	Canada	France	W. Germany	Italy	Netherlands	U.K.	Sweden	Japan
1. Energy embodied in nonenergy imports	135.2	46.9	69.6	96.3	42.0	40.2	63.3	19.8	43.7
2. Energy embodied in nonenergy exports	122.3	49.2	67.6	128.9	47.8	40.6	66.4	25.1	83.0
3. Net energy embodied in nonenergy trade	+12.9	−2.3	+2.0	−32.6	−5.8	−0.4	−3.1	−5.3	−39.3
4. Part of refinery losses attributed to trade in products	+10.3	+0.1	−0.3	+2.5	−1.2	−1.9	+0.2	+0.9	+0.9
5. Total energy embodied in nonenergy trade in petroleum products (3 + 4)	+23.2	−2.2	+1.7	−30.1	−7.0	−2.3	−2.9	−4.4	−38.4
6. As percent of total energy consumption	+1.3	−1.2	+1.0	−11.9	−5.3	−3.7	−1.4	−10.2	−12.5

Note: Figures with + signs indicate import surpluses; those with − signs indicate export surpluses.
Source: RFF estimates.

These are of course very rough-and-ready estimates. The trade data are expressed in U.S. dollars converted at the market rate of exchange, a procedure found to be generally unsatisfactory due to the existence of differences in relative prices between countries. In addition, the application of input–output-derived U.S. energy coefficients to foreign trade components of all our other countries is open to objection. On the other hand, the results agree very closely with studies recently available for several industrial countries based on own-country coefficients.[27] This similarity in results for countries for which independent estimates are available gives grounds for confidence in the estimates for the other countries.

An additional form of embodied energy that has been touched upon earlier but not quantified is the amount of energy saved if fuels are imported in a finished rather than a crude state, thus avoiding the consumption of energy that would have been needed to upgrade them. Similarly, the export of finished fuel products embodies the energy used up in converting them from the crude state. This form of energy consumption appears in our balance sheets under consumption by the energy industry, which, it will be recalled, includes the most energy-intensive industries of them all. The most important example of this trade is petroleum refining, although there is in addition some limited trade in coke.

Although all countries have some trade in petroleum products, in many cases the net balance of such trade is small in relation to refinery output. And for those countries whose net trade in products is large in relation to refinery output, refinery losses (that is, the amount of energy used up in the refinery operation) are small in relation to total energy consumption. In consequence, for most of these countries even a substantial adjustment to refinery losses to take into account a net gain or loss from trade in

[27] For United States N. S. Fieleke, "The Energy Trade: The United States in Deficit," New England Economic Review (May/June 1975). This article reports a small net export in nonenergy trade rather than the small net import given in our results. See also Robert A. Herendeen and Clark W. Bullard III, "U.S. Energy Balance of Trade, 1963–73," mimeo. (Urbana, Ill., University of Illinois, 1974); Alan Strout (associated with the Energy Laboratory at the Massachusetts Institute of Technology in Cambridge, Mass.), "The Hidden World Trade in Energy," mimeo., 1974. For Canada R. P. Gupta, "Energy, Labor, Capital and Primary Input Requirements of Industries, Goods and Services in Canada," mimeo. (Winnipeg, The Biomass Institute, 1975). For West Germany Richard V. Denton, "The Energy Costs of Goods and Services in F. R. of Germany," Energy Policy vol. 3, no. 4 (December 1975). For United Kingdom David J. Wright, "Goods and Services: An Input Output Analysis," Energy Policy vol. 2, no. 4 (December 1974); this article reports a small net import in place of our small net export. For Sweden Industridepartementet, Energy—1985, 2000 (Stockholm, Statens Offentliga Utredningar, 1974).

products would not affect total energy consumption significantly. The two exceptions are the Netherlands and Sweden. The Netherlands exports about one-half of its refinery output so that one-half of its refinery losses, or about 3 percent of total energy consumption, represents embodied energy in fuel exports. Sweden, on the other hand, imports almost three times as many products as it refines domestically.[28] If these imported products had been refined domestically, refinery losses would have been three times higher, representing over 2 percent of total energy consumption. A similar exercise could be done for trade in coking coal contrasted with coking domestically. In practice, the quantities involved are very small. Only Sweden imports a major part of coke supplies (75 percent), and even in this case losses are a negligible part of total energy consumption.

Taking these two elements together—energy embodied in nonenergy foreign trade and energy embodied in trade in energy products—the following picture emerges. For the United States, Canada, France, and the United Kingdom, embodied energy flows are more or less in balance, without any significant inflow or outflow. All the other countries have a net outflow. In the case of Italy and the Netherlands this amounts to between 4 and 5 percent of total energy consumption. For Germany, Japan, and Sweden it amounts to between 10 and 13 percent. Overall, embodied energy, whether in nonenergy trade or in trade in energy products, widens the difference in consumption levels between the United States and other countries. These findings are consistent with the findings of the input–output analysis in chapter 7.

Influence of Fuel Mix in the Industrial Sector

As in the household-commercial sector, the industrial sectors of our countries consume a wide variety of fuels in very different proportions. For most countries, petroleum products are the single most important fuel. For the United States and the Netherlands, however, natural gas is the largest single fuel source, providing over 50 percent of the total. Use of coal by industry varies substantially among countries; the comparative proportions of the total provided by electricity also differ but less sharply.

This diversity in source of fuel raises the question of whether the fuel mix influences the level of consumption through the disproportionate use of less efficient fuels in some countries.

---

[28] The construction of a recent new refinery in Sweden may have changed this position significantly.

The thermal efficiency of fuels in industrial uses varies considerably. Coal is usually considered to have a thermal efficiency of about 70 percent; petroleum products, 80 percent; gas, 85 percent; and electricity, 100 percent.[29] Countries with a large proportion of coal in their fuel mix and a low proportion of electricity would consequently be expected to use more energy to produce the same amount of useful energy as those countries whose fuel mix is concentrated on the higher efficiency fuels.

In order to see whether varying efficiencies of fuel mix had any effect on consumption levels, the quantities of each fuel consumed by industry were multiplied by the relevant efficiency to arrive at a sum representing more closely the "useful energy" to be obtained from a given amount of energy input.

The results show that despite the considerable variation in fuel mix among European countries, the proportion of useful energy derived from total inputs is very close—about 60 percent. The United States is slightly higher than most countries, at 65 percent, due to a high proportion of gas and electricity in the industrial fuel mix, which compensates for the high proportion of coal. Thus the fuel mix does affect the amount of useful energy obtained, but not in a way that would help explain the generally higher U.S. industrial energy consumption relative to GDP.

It must be emphasized that the above adjustment relates to *thermal* efficiency only, and as such, obscures or underestimates other properties of the various fuels. Coal, for example, is needed in iron making because of its solid properties; other fuels would not serve. Thus, countries with a larger iron and steel sector will tend to have a large proportion of total fuel in the form of coal, which translates, according to the above guideline, into a relatively low thermal efficiency.

Transformation Losses

Although transformation losses are conceptually and, in this study, statistically, distinct from energy consumption by the individual consuming

[29] F. Gerard Adams and James M. Griffin, "Energy and Fuel Substitution Elasticities: Results for an International Cross Section Study," mimeo. (Philadelphia, Pa., Economics Research Unit of the University of Pennsylvania, 1973). Note that the electricity efficiency does not include heat losses. If heat losses were included, electricity efficiency would fall sharply; and William D. Nordhaus, ed., *Proceedings of the Workshop on Energy Demand* (Laxenburg, Austria, International Institute for Applied Systems Analysis, May 1975).

Note that coal use is not invariably less efficient than oil and gas use in industry. For process steam generation, oil- and gas-fired package boilers are widely used in the United States. These low-cost devices have tended to be less efficient than the more costly field-erected boilers needed for coal use.

sectors, they have been included at the end of this chapter on industry, rather than anywhere else, because the most important transformation activities are essentially industrial in nature.

Excluding for the moment Canada and Sweden, whose transformation losses appear very high due to the hypothetical losses assigned to their large hydroelectric generation (see chapter 2), the transformation losses of most countries are closely grouped around 15 to 17 percent of total energy consumption. The exceptions are the United Kingdom—high at 23 percent—and the Netherlands, low at 13 percent. For many countries, the size of transformation losses is comparable to total energy consumed in, say, the transportation sector. The U.S. figure is 17 percent.

While transformation losses in most of the countries are similar in magnitude to energy consumption in transportation, they account for a much smaller part of the total difference in overall energy/GDP ratios— averaging 15 percent, compared with an average of 45 percent accounted for by the transport sector. Their contribution to the total difference in energy/GDP ratios is comparable—though rather smaller—to that of the industrial sector (including energy) and the household-commercial sector.

Transformation losses are losses sustained within a wide range of activities, including patent fuel plants (which manufacture coal briquette), coke ovens, gas works, and electricity generation. It has not been possible to identify the energy inputs and outputs of all these activities, nor indeed is such an effort worthwhile for our purposes, as many of them incur negligible losses. We therefore limited ourselves to the losses incurred in electricity generation, which account in all cases for over 90 percent of total losses.

Considering electricity generation as a producing industry, rather than as a source of power, it would be of interest to see how much of the difference in heat losses relative to GDP is due to differences in the efficiency with which electricity is generated in the various countries and how much is due to the relative size of the electricity-generating capacity (see table 6-14). We found, however, that detailed data on electricity generation was not good enough to draw confident conclusions about differences in heat rates, and we therefore standardized thermal efficiencies on a basis of 35 percent gross (see chapter 2).

The standardization of thermal efficiencies in conventional generating sets may not violate reality, because, with one or two exceptions, the thermal efficiencies of most countries fall within a fairly narrow range. Of greater importance is the fact that thermal efficiencies calculated as

TABLE 6-14 Differences in Transformation Losses in Electricity Generation Between Countries, 1972

Country	Transformation losses relative to GDP		Difference from U.S. in transformation losses relative to GDP (tons oil equiv. per $ million GDP)	Transformation losses as a percentage of total difference in energy/GDP ratios
	Tons oil equiv. per $ million GDP	U.S. = 100		
U.S.	250	100.0	—	—
Canada	401	160.4	−151	51.7
France	140	56.0	110	16.1
W. Germany	170	68.0	80	17.8
Italy	133	53.2	117	20.7
Netherlands	164	65.6	86	41.3
U.K.	267	106.8	−17	−1.1
Sweden	254	101.6	−4	−4.1
Japan	147	58.8	103	16.3

Note: Blanks = not applicable.
Source: Data in appendix B.

electricity output in terms of fuel input do not take into account co-generation, that is, generating sets usually operated by industrial producers that are designed to use both electricity *and* heat. (In our format, such heat would appear within transformation losses.) Including heat use, such plants achieve higher effective thermal efficiencies than conventional generating systems. This is likely to be of relevance to a comparison between the United States on the one hand and countries like Sweden and West Germany, where co-generation is fairly important, on the other.[30]

The standardized efficiencies used here, however, mean that by definition, differences in transformation losses relative to GDP (table 6-15) are due entirely to differences in relative size of the generating capacity. Leaving aside Canada and Sweden, the other countries on average consume about 40 percent less than the United States, but with some variation among countries. France, for example, consumes less than half the electricity (relative to GDP) than the United States. The United Kingdom, on the other hand, consumes about 20 percent less.

[30] See Stanford Research Institute, *Comparison of Energy;* and Schipper and Lichtenberg, *Efficient Energy.*

TABLE 6-15   Heat Losses and Electricity Generated, 1972
(tons oil equivalent per $ million GDP)

	Heat losses[a]	Electricity generated	Index of heat losses and electricity generated U.S. = 100
U.S.	268	144	100
Canada	367	198	137
France	123	66	46
W. Germany	186	100	69
Italy	152	82	57
Netherlands	157	84	59
U.K.	222	120	83
Sweden	287	154	107
Japan	189	102	71

Source: Data in appendix B.
[a] Based on a standardized efficiency of 35 percent.

Despite the generally unsatisfactory data on heat rates, two observations can be made with some confidence. The first is that the United Kingdom has a substantially lower thermal efficiency than that of the other countries.[31] This appears to be due in part to older generating sets. Comparing the United Kingdom with other European countries, therefore, suggests that the higher U.K. transformation losses relative to GDP are due both to a larger electricity sector (relative to GDP) and to higher heat rates. A comparison between the United States and the United Kingdom, on the other hand, indicates that all and more of the slightly greater U.K. heat losses relative to GDP are due to the lower U.K. heat rates, as the United States has a relatively larger electricity sector.

Second, even though heat loss data are generally unsatisfactory for our countries as a group, for some individual countries they appear to be more satisfactory. Thus, for the United States, France, and West Germany, unstandardized data give similar and reasonable heat rates—about 35 percent. Sweden appears to be somewhat higher. In this event, almost all of the differences in heat losses relative to GDP between the United States, France, and West Germany would be due to the larger U.S. electricity sector.

[31] The European Communities' *Energy Statistics Yearbook, 1969–72,* gives an average specific consumption of 375 coal equivalent grams per kWh gross, compared with 333 for West Germany, 316 for France, 315 for Italy, and 326 for the Netherlands.

Summary

Our nine-country survey of industrial energy consumption discloses that at the level of aggregation examined, the comparatively high ratio of U.S. industrial energy consumption to national GDP arises from high U.S. industrial energy intensities. It does not appear to be due to structural factors, since both industry as a whole and several specific industrial sectors for which we developed multicountry data, represent disproportionately low shares of aggregate GDP in the United States. For example, the United States has a small iron and steel industry relative to other countries but a higher consumption of energy per ton of crude steel. This high energy intensity appears to be due to a relatively high proportion of the high-energy-using open-hearth process and a relatively low proportion of continuous rolling capacity.

We noted that differences in energy intensities need not reflect differences in *overall* efficiency. Empirical evidence on the extent to which energy intensity relates to economic efficiency is beyond the scope of our report. Also, the analysis apportioning structural and intensity factors is a function of the level of aggregation being considered. Successive disaggregation beneath the levels of detail to which we were compelled to limit ourselves can always reveal a different apportionment of these two elements. We noted several examples of this phenomenon in some of the comparisons.

In spite of these qualifications, the evidence of pervasively lower foreign industrial energy intensities is sufficiently compelling to justify efforts to probe their economic attractiveness and, beyond that, to evaluate their potential applicability to the United States (and Canadian) setting. The co-generation of industrial steam and electric power, and the place for waste heat recovery methods, illustrate industrial energy systems whose economic benefits deserve attention.

PART III

Input–Output and Final-demand Analysis

CHAPTER 7

Energy Embodied in Components of Final Demand

The Two Approaches Contrasted

IT SEEMS APPROPRIATE at this point to clarify the differences in method between the input–output approach of this chapter and that employed in the sectoral discussion in chapters 4 through 6. The latter is based on the OECD energy balance tables for the countries included in the study and involves a sector-by-sector analysis of energy consumption in which the sectors are broadly defined by function. Based on these broad definitions, the sectoral approach made it possible to analyze, in varying levels of detail, the way in which countries differ in their use of energy for broadly similar purposes and how these differences might affect the disparities in overall energy/GDP ratios. In addition, to the extent that the detailed energy-use data within sectors and supplementary information permitted, estimates were developed of how much of inter-country differences in sectoral energy/output ratios were due to: (1) structure (composition of output) and (2) the energy intensity of indi-vidual activities within each sector. As noted earlier in the study, this aspect is a major element in our analysis of intercountry differences in energy/GDP ratios.

However, the structure–intensity analyses, based on the energy balance table sector approach, were limited, with the analyses not entirely con-sistent from sector to sector. More serious, however, was the fact that the estimates for the various sectors were, strictly speaking, not additive, be-cause the output measures used in deriving the intensity factors were not based on a common unit of measure. Consequently, only approximate estimates of differences in output mix versus differences in energy intensi-ties of the individual activities could be made.

143

In contrast, the input–output approach to the mix versus energy intensity analysis is comprehensive, systematic, and can be used to factor the intercountry differences in energy/GDP ratios into the two elements of mix and energy intensity.[1] This approach uses detailed information on the composition of final goods and services produced by each country, and an input–output table for the United States, to determine how much of the total difference between the energy/GDP ratios of other countries relative to the United States is due to (1) the differences in the mix of final goods and services and (2) other factors, primarily differences in total energy (direct and indirect) embodied in the individual final products and services.

Input–output tables provide detailed information on what each industry buys from every industry in the economy and what, in turn, it sells to every industry in the economy, including sales of goods and services to final demand (the final expenditure side of the national income and product accounts). This information is used to derive input–output direct coefficients, that is, for a given industry, the amount in value terms of each input purchased from other industries, as well as from itself, for each dollar of industry output. Examples of direct coefficients as they relate to energy analyses are inputs purchased from each of the energy industries per dollar of output of the automobile industry. Since the automobile industry also uses steel, aluminum, glass, and the like, to produce automobiles, and each of these industries in turn requires energy to produce its products, it is possible to use all of these direct coefficients to derive a cumulative total of how much output is required, directly or indirectly, from each energy industry per dollar of automobile output. Such estimates of total requirements from each industry per dollar of output of each end product or service are calculated as part of a set of input–output tables. The total requirement coefficients include not only changes in the direct energy coefficients but all other coefficients—for example, substitution of energy-intensive aluminum for less energy-intensive materials in the production of automobiles. Such changes may, in turn, affect energy requirements.

It is possible, in concept, to use input–output tables of various countries to determine how much of the difference in energy/GDP ratios is due to the differences in the patterns of final demand among countries and how

---

[1] The terms "mix" and "energy intensity" will vary depending on the analytical approach, that is, on whether it is the input–output approach of this chapter or the sector-by-sector approach used earlier. For further discussion of these differences, see footnote 1 in appendix G, and chapter 9.

much is due to differences in production, transportation, and distribution technologies, as reflected in their respective input–output coefficients. This type of intercountry comparison based on input–output tables requires both detailed estimates of final demand and input–output tables that are comparable between countries in terms of a common set of prices, industry classification systems, and accounting conventions.

Unfortunately, the input–output tables of the other countries in our study are not directly comparable to the U.S. tables, either in terms of a common set of prices, industry classification systems, or accounting conventions. To avoid this difficulty, an alternative procedure was used in this report. We used U.S. input–output tables, plus final-demand expenditures for other countries, which were derived from data in the *ICP Report*,[2] to determine how much of the difference in energy/GDP ratios of other countries relative to the United States might be due to the difference in the composition of final-demand expenditures—with the remaining difference presumed to be due primarily to differences in input–output coefficients (as they affect total energy intensity coefficients). The details of this approach and analysis of the results are the subject of this chapter. A more detailed statement on methodology is in appendix G.

Methodology of the Input–Output, Final-demand Analysis

As noted above, the basic procedure in input–output analysis involves the use of final-demand "bills of goods" and input–output total requirement coefficients to derive estimates of direct and indirect output required from each industry. The *ICP Report,* used in this study as the basis for deriving the GDP part of the energy/GDP ratios, also provides detailed estimates of final-demand expenditures for two major components: consumption and fixed capital expenditures, which together account for about 85 to 90 percent of total GDP for the countries included in both the ICP comparison and this study (United States, France, West Germany, Italy, United Kingdom, and Japan). Only aggregate estimates for the other components of GDP, that is, inventory change and net exports, and government purchases are given in the *ICP Report.* As a consequence, the input–output approach was used for the consumption and fixed investment components but not for the government and "other" compo-

---

2 Irving B. Kravis, Zoltan Kenessey, Alan Heston, and Robert Summers, *A System of International Comparisons of Gross Product and Purchasing Power* (Baltimore, published for the World Bank by The Johns Hopkins University Press, 1975).

nents. Estimates for the government and other components were developed as a group, based on a more aggregative approach than that used in the input–output analysis. (This approach will be described later.)

The development of consumption and investment bills of goods involved several types of adjustments in order to make the final-demand estimates consistent with the input–output table used for this analysis. The basic data as given in the *ICP Report* were: (1) in each country's own currencies and prices, (2) in purchasers' prices, that is, the prices paid by the consumer rather than the producer (the difference being primarily transportation and distribution charges), and (3) classified by functional categories rather than by producing industry. In order to assure consistency with the U.S. input–output table, the basic data were first converted to U.S. purchasers' prices based on the detailed purchasing power parities given in the *ICP Report* and then, by use of a U.S. bridge table, were converted into producers' prices and classified by producing industry.

This procedure involved grouping the original 109 ICP consumption items into 61 product groups and grouping 14 construction types into 4 types. The 22 producer-durable equipment items matched the bridge table and were used in full detail. The purpose of the bridge table was to distribute the functional groups or activities into producing-industry categories, including allocation of the spread between producers' and purchasers' prices to the various transportation and distributor industries. This meant that the U.S. patterns of transportation and distribution and the U.S. industry mix of production were both assumed in distributing each of the initial 87 product groups or activities into end product deliveries classified on an input–output, producing-industry basis.

There is one other aspect of the input–output analysis that should be mentioned since it also influences the interpretation of the effect of differences in final-demand mix on energy/GDP ratios. This is the particular form of the input–output table used. The U.S. table is based on the official 1963 input–output table developed by the Bureau of Economic Analysis in the Department of Commerce. The table was aggregated to 135 industries and updated to 1970 by the Bureau of Labor Statistics. The particular form of the table is based on the concept of total supply as the measure of industry "output" (including imports, which are considered to be comparable to the domestic products). This means that the use of such a table to convert final demand for goods and services in other countries into direct and indirect energy consumption assumes not only U.S. technology as reflected in U.S. input–output coefficients, but also

U.S. proportions of domestic supply and imports in meeting requirements for materials such as steel. This in turn affects total energy requirements, since it excludes the energy used in other countries to make the products that are imported and used in the United States.

In the final stage of the procedure, multiplying the consumption and investment bills of goods against the U.S. total coefficient table yielded estimates of energy consumption stated in terms of dollars of output required from the various energy-producing industries. To determine how much of the difference in the total energy/GDP ratios was due to the difference in consumption and investment final-demand mix, the generated energy consumption requirements were converted into thermal units such as the million tons of oil equivalent (mtoe) used in the initial energy/GDP ratios of earlier chapters. This was done as part of the estimating procedure, based on conversion factors for each energy-producing industry. In *this* final-demand analysis, however, the dollars were converted into British thermal units (Btus) so as to be consistent with the Bureau of Mines' estimates of U.S. energy consumption by type of energy product for the year 1970. Since both mtoes and Btus are thermal measures of energy, and the results of the input–output exercise are expressed as Btu/GDP ratios, the results can be compared to the overall mtoe/GDP ratios.[3] The method for the conversion of energy consumption from dollars to Btus is described in appendix G. Briefly, it involved a two-stage process in which initial estimates of energy consumption were based on the assumption that all users of each type of energy (electric power, gas, coal, and the like) pay the same average price, that is, the Btu content per dollar of purchase is the same for all users of a particular type of energy. These estimates were then modified by a set of adjustments to take account of the differential in average price paid by household users of energy and other users of energy (primarily industrial and commercial users).

Because the assumption that all users of each type of energy pay the same price (except for the differential between direct purchases of energy by consumers and "all other" users) holds only at very broad levels of aggregation, the conversion from dollars to thermal units was performed only for major categories of final demand, and the results are shown only at these broad levels.

---

[3] For simplicity in general discussions of the ratio of energy consumption to the gross domestic product, the term energy/GDP may be used in lieu of mtoe or Btu/GDP.

To summarize, the input–output analysis gives us estimates of differences in energy per dollar ratios in other countries relative to United States based on differences in patterns of consumption and fixed capital expenditures, distributed among 87 product groups or activities, and stated in purchasers' value. The estimates do *not* reflect disparities in total domestic energy required per dollar of final demand within each of the product groups, but only differences among the product groups. Total energy intensity for each product group includes energy used in all stages of production, transportation, and distribution. Energy consumption was converted from dollars to thermal units based on the assumption that all users pay the same price for each type of energy, adjusted for differential prices paid by household users relative to other users. Because the assumption holds only at broad levels of aggregation, the conversions were made only for major categories of final demand.

Proportion of Total Final Demand Covered by Input–Output Analysis

Since the input–output analysis was limited to the consumption and fixed investment components of GDP, it is important to see what proportion of total GDP was covered by the input–output analysis. The percentage distribution of GDP and its major components, based on the estimates in the ICP study, is shown in table 7-1. It should be noted that the ICP estimates for consumption and fixed investment differ in coverage from the similar categories in the U.S. official national accounts because these categories in the U.S. accounts are limited to the private sector, whereas the ICP estimates for consumption and investment include certain types of government expenditures: specifically, government expenditures on health, education, and recreation are included in consumption expenditures; and nondefense government expenditures for construction and producer-durable equipment are included in the investment components.

The components of GDP included in our input–output analysis represent by far the largest share of GDP in all countries, ranging from 85 to 91 percent of the total, with United States at the lower end of the range. The lower proportion of GDP in United States accounted for by consumption and fixed investment is offset by government purchases, which account for almost 15 percent of total GDP in United States compared to 7 to 12 percent in the other countries. The remaining components not included in the input–output analysis are, on the whole, a small proportion of total GDP, but this may be misleading in terms of the energy/GDP

TABLE 7-1 Percentage Distribution of GDP and its Major Components Among Six Countries, 1970

GDP and components	U.S.	France	W. Germany	Italy	U.K.	Japan
Components included in I-O[a] analysis						
Consumption	68.2	63.9	57.3	70.1	70.3	56.4
Fixed investment	16.9	26.6	29.4	21.2	17.0	28.6
Subtotal	85.1	90.5	86.7	91.3	87.3	85.0
Other components						
Net inventory change	0.4	2.3	1.6	1.1	0.5	3.0
Net exports	0.2	0.2	1.5	0.1	0.5	0.8
Government	14.4	7.1	10.3	7.6	11.7	11.3
Subtotal	15.0	9.6	13.4	8.8	12.7	15.1
Gross domestic product	100.0	100.0	100.0	100.0	100.0	100.0

Note: Detail may not add to 100 due to rounding. Percentage distribution based on expenditures in U.S. prices.
[a] Input–output.

comparisons for the net export category. It represents, of course, the net balance between imports and exports and may be small in terms of the net result, but gross exports and imports may be quite large and, depending on differences in energy intensity of imports and exports in relation to each other and to the other components of final demand, may contribute more to the differences in the energy/GDP ratios than the small size of the net export figures would suggest. This topic is also discussed in chapter 6 of this study.

Results of the Input–Output Analysis

Although the actual computations were done in considerable detail (see appendix G), the results can be summarized in terms of the two major components, consumption and fixed investment, and major categories within each. For consumption, there are three categories; two energy categories covering (1) fuel and power for household operations and (2) gasoline, oil, and grease primarily for personal automobile transportation. The third consumption category covers all other consumer purchases. Fixed investment included two components: construction and producer-durable equipment.

It will be recalled that the energy consumption estimates include both direct and indirect energy used in producing, transporting, and delivering each of the energy types to the final-demand consumer.[4] Therefore, the figures shown for the energy content of consumer purchases of energy in this part of the study are higher than similar figures for such purchases in other chapters of the study. The latter figures generally exclude energy used or lost by the energy industries themselves in the conversion, processing, and transportation of energy.

Finally, unless otherwise stated, the term final demand will be used in this section of the chapter to include the components of consumption and fixed investment only.

With this as background, what conclusions emerge from this analysis? First, the estimates as shown in table 7-2 and particularly the last line in the table indicate that the composition of final demand in other countries is sufficiently different from the U.S. mix that, given U.S. input–output coefficients, U.S. bridge tables, and the like, along with the differences in the level of energy intensities of these end items, the effect is to reduce the aggregate Btu/dollar ratios for other countries by 14 to 26 percent below the aggregate U.S. ratio for the same components, with the European countries averaging about 20 percent and Japan 23 percent below the U.S. ratio.

Analysis of the indexes for individual categories suggests that differences in mix within categories, with the exception of purchases for consumption of fuel and power, were not a major factor in reducing the aggregate Btu/dollar ratio below the U.S. level. On the contrary, the indexes for consumer nonenergy purchases, construction, and producer-durable equipment are about the same as those of the United States or higher. The indexes for purchases of fuel and power for household operations, on the other hand, are substantially lower for all the other countries relative to United States. The difference in the mix of types of energy purchased by households in other countries is in the direction of less energy-intensive types of fuels per dollar of such purchases than in United States. Other countries covered in this comparison, for example, use relatively less gas than we do in United States, and gas has a much higher Btu per dollar ratio than the other forms of energy.

The lower Btu/dollar ratio for the fuel and power category as a whole, in other countries relative to United States, is more than offset in its effect on total direct energy use by consumers by two factors: (1) that energy

___

[4] The energy content of electricity, for example, includes the heat loss in electric power generation, which is about two-thirds of the total primary energy content of delivered electric power. Similarly, the energy content of gasoline includes not only the energy content of crude oil used to make gasoline but the energy used to refine the crude oil.

TABLE 7-2 Effect of Final-demand Mix on Component Btu/Dollar Ratios (Consumption and Fixed Investment Components)

Final-demand components	U.S.	France	W. Germany	Italy	U.K.	Japan
			billion 1970 dollars			
Consumption	665.0	126.8	133.1	97.9	126.7	189.9
Energy purchases	44.6	3.1	4.7	2.1	5.3	3.0
Fuel and power	21.6	2.2	3.2	1.5	3.9	2.5
Gasoline, oil, grease	23.1	0.9	1.4	0.6	1.4	0.4
Nonenergy purchases	620.4	123.8	128.4	95.8	121.4	186.9
Fixed investment	166.6	53.1	69.1	29.7	31.0	96.9
Construction	95.1	34.9	46.3	21.8	18.3	61.8
Producer durables	71.5	18.2	22.7	7.9	12.7	35.1
Total purchases	831.6	180.0	202.1	127.6	157.7	286.8
			percentage distribution			
Consumption	80.0	70.5	65.8	76.7	80.3	66.2
Energy purchases	5.4	1.7	2.3	1.6	3.3	1.0
Fuel and power	2.6	1.2	1.6	1.2	2.5	0.9
Gasoline, oil, grease	2.8	0.5	0.7	0.5	0.9	0.1
Nonenergy purchases	74.6	68.8	63.5	75.1	77.0	65.2
Fixed investment	20.0	29.5	34.1	23.3	19.7	33.8
Construction	11.4	19.4	22.9	17.1	11.6	21.6
Producer durables	8.6	10.1	11.2	6.2	8.1	12.2
Total purchases	100.0	100.0	100.0	100.0	100.0	100.0
			quadrillion Btus			
Consumption	50.3	7.0	7.7	5.0	7.9	10.0
Energy purchases	25.9	1.9	2.7	1.3	3.1	1.8
Fuel and power	15.1	1.5	2.0	1.0	2.5	1.6
Gasoline, oil, grease	10.9	0.4	0.7	0.3	0.6	0.2
Nonenergy purchases	24.3	5.1	5.0	3.7	4.8	8.3
Fixed investment	9.5	3.1	4.1	1.8	1.8	5.9
Construction	6.1	2.2	3.0	1.4	1.2	4.2
Producer durables	3.4	0.9	1.1	0.4	0.6	1.7
Total purchases	59.8	10.1	11.8	6.8	9.7	15.9
			thousand Btus per dollar, ratio			
Consumption	75.6	55.0	57.8	49.6	62.3	52.9
Energy purchases	581.2	627.2	582.2	602.0	588.0	594.6
Fuel and power	698.5	692.7	631.3	654.6	628.1	614.5
Gasoline, oil, grease	471.7	471.7	471.7	471.7	471.7	471.7
Nonenergy purchases	39.2	40.9	38.7	38.6	40.0	44.3
Fixed investment	57.0	58.6	59.7	60.6	58.3	61.0
Construction	64.0	63.0	64.8	63.7	63.7	67.4
Producer durables	47.8	50.2	49.3	51.9	50.5	49.8
Total purchases	71.9	56.1	58.5	52.9	61.5	55.6
			index of ratio, United States = 100			
Consumption	100.0	72.8	76.5	65.6	82.4	69.9
Energy purchases	100.0	107.9	100.2	103.6	101.2	102.3
Fuel and power	100.0	99.2	90.4	93.7	89.9	88.0
Gasoline, oil, grease	100.0	100.0	100.0	100.0	100.0	100.0
Nonenergy purchases	100.0	104.3	98.7	98.3	100.8	112.9
Fixed investment	100.0	102.8	104.7	106.2	102.3	107.0
Construction	100.0	98.5	101.2	99.6	99.5	105.3
Producer durables	100.0	105.1	103.3	108.6	105.7	104.3
Total purchases	100.0	78.0	81.3	73.6	85.6	77.4
Difference from U.S.	—	−22.0	−18.7	−26.4	−14.4	−22.6

Notes: Detail may not add to total due to rounding. Blank = not applicable.

purchases for household operation are relatively more than gasoline purchases in other countries compared to the U.S. and (2) that the Btu content of the fuel and power category per dollar of purchases is more than the Btu/dollar ratio for gasoline. The result is that for direct energy purchases as a whole, the Btu/dollar ratio is higher in most of the other countries than in United States, in spite of the fact that for the two components the ratios are either lower (fuel and power) or the same (gasoline).

If we look at the Btu/dollar ratios at a higher level of aggregation namely, at the level of the two major components (consumption and fixed investment), we find very different patterns for the two. In the case of fixed investment, the indexes of the Btu/dollar ratios for other countries are somewhat higher than that for the United States and not too far out of line with the indexes for the two component categories—construction and producer durables (which are either slightly below or somewhat higher than the ratios for the United States).

In contrast, the Btu/dollar ratios for total consumption are all much lower than that for United States, averaging about 25 percent lower for the European countries and 30 percent lower for Japan. This pattern of lower Btu/dollar ratios for total consumption is substantially different from the Btu/dollar ratio for the two component categories—total consumer energy purchases and all other consumer purchases—which are at about the same level as the U.S. ratio, or somewhat higher. Here again the explanation for this seeming anomaly lies in the combined effect of the difference in the relative importance of the two component categories and the substantial differences in the relative levels in Btu/dollar ratios for the two categories. In this case, we find that other countries devote a much lower proportion of total consumption expenditures to energy purchases, which have a much higher Btu/dollar ratio, than they do to nonenergy purchases. Put another way, the aggregate Btu/dollar ratio for total consumption expenditures in the United States is very much higher than that for the other countries, not because the average Btu/dollar ratios for the two component groups (energy and all other consumer purchases) are that much different from the comparable ratios in the other countries, but because (1) U.S. consumers devote proportionately more of their budgets to energy purchases and (2) because the direct purchases of energy by consumers is so much more energy intensive than the average energy intensity of all the other nonenergy consumer purchases.

The last element of the mix effect to be discussed is the one between the two major components—consumption and fixed investment. Again, the mix effect of the two components will depend on the difference in the proportion of total final demand accounted for by these components, relative to the United States, and the difference in their energy intensities. In the case of fixed investment, the Btu/dollar ratio for the component is only somewhat higher than that in the United States, but the proportion of total final demand going into fixed investment is very much higher in the other countries, with the exception of United Kingdom. The proportion is 20 percent in United States; it averages about 30 percent for the other countries, except for United Kingdom, which has about the same percentage devoted to fixed investment as does United States. The net effect of these two factors tends to increase the aggregate Btu/dollar ratios in other countries relative to United States.

The opposite situation holds for consumer expenditures. Here both factors, energy intensity ratios and the relative importance of such expenditures in total final demand, are in general substantially lower than in United States and reinforce each other, resulting in contributing to lower Btu/dollar ratios in other countries relative to United States.

Derivation of Final-demand Mix Effects on Total Energy/GDP Ratios

Up to this point, the analysis of the effect of differences in final-demand mix on Btu/dollar ratios has been limited to the components included in the input–output analysis, that is, consumption and fixed capital expenditures. As noted earlier, in 1970 these components account for about 85 percent of total GDP in the United States and about 85 to 91 percent in the other countries.

Estimates of Remainder of Final-demand Components

In order to complete the analyses of the effect of differences in final-demand mix on overall energy/GDP ratios, it is necessary to provide estimates for the remainder of GDP that covers government purchases of goods and services, net exports, and net inventory change. Unfortunately, the *ICP Report* gives only aggregate data for the remaining three components, except for government purchases, where the distribution between government purchases of total commodities and government employee compensation is given. This latter distribution is quite important in assessing the effect of final-demand mix on differences in energy/GDP ratios, because compensation of government employees by definition does not involve direct or indirect energy consumption but accounts for a significant proportion of total GDP (about 7 percent in the United States, and 4 to 9 percent in the other countries).

From the limited information in the *ICP Report* on government purchases and the other components of final demand, we derived estimates of energy consumption and energy-intensity ratios, based on an aggregate approach that groups the remaining final-demand expenditures into two categories: (1) compensation of government employees and (2) government purchases of commodities, net exports, and net inventory change. This latter "residual" category accounts for about 8 percent of GDP in the United States and 4 to 9 percent in the other countries, roughly the same general proportions of GDP as government compensation (although not necessarily for each country).

The procedure used was to derive a residual estimate of energy consumption in the United States for the remaining final-demand components. We did this by deducting energy consumption previously estimated in the input–output analysis for consumption and fixed capital expenditures from the figure for total domestic energy consumption in 1970. The residual energy consumption estimate, when related to the residual final-demand estimate (excluding government compensation), yielded an energy-intensity ratio (Btu/dollar) for the group. This ratio could then be used, along with the zero energy-intensity ratio for government compensation, to derive a weighted average energy-intensity ratio for the remainder of final demand in each country, with the average varying depending on the relative importance of government compensation and the "all other" category. The results of the calculations are shown in table 7-3. These were then combined with aggregate estimates for the major components included in the input–output analyses to yield estimates of the effect of differences in final-demand mix for all GDP components on total energy/GDP ratios for other countries relative to the United States.

The derived average Btu/dollar ratios for the remaining group of final-demand components not included in the input–output analysis are substantially lower than the comparable ratios for the other major components of final demand. This is largely due to the dampening effect on the average of the zero energy content of government payrolls, since the Btu/dollar ratio of 57,200 for the remainder of this group (that is, the commodity portion of government purchases, net exports, and net inventory change) is, with the exception of the United States, about the same as the average Btu/dollar ratio in other countries for total consumption and fixed capital expenditures combined. The inclusion of government compensation reduces the average Btu/dollar ratio for the group substantially for all countries, but the resulting lower ratios do not fall into any uniform pattern relative to the United States, with some being higher and others lower.

Effect of Total Final-demand Mix on Btu/GDP Ratios

When the estimates of Btu/dollar ratios for the remaining final-demand components, taken as a whole, were combined with those for the components covered in the input–output analyses—taking account of the relative importance of the two groups in each country—the net effect of final-demand mix on overall Btu/GDP ratios was derived, relative to the United States. In general, the result of adding the estimates for the residual group of government purchases, and the like, to the previous analysis was either to reduce the impact of final-demand mix (increase the ratios relative to the United States) or leave them about the same. Put another way, the result of including the residual group in the mix analysis reduces the Btu/dollar ratios for all countries below that based on the consumption and fixed investment components alone, but the reduction is relatively more in the case of the United States than for three of the other countries, and about the same for the other two.

The final estimates of the effect of differences in mix of all final-demand components show that overall Btu/GDP ratios in other countries are almost 19 percent below the ratio for the United States, on the average. The spread among the ratios is fairly wide, ranging from -14 percent in the case of United Kingdom to -23 percent for Italy and Japan. In between are France, at -18 and Germany at -15, both below the unweighted average of the five countries relative to the United States.

Contribution of GDP Components to the Final-demand Mix Effect on Btu/GDP Ratios

We now examine the extent to which the specific components of final demand contribute to the differences in the overall Btu/GDP ratios. The contribution of each component is a function of the average energy intensity of the component and its relative importance in total final demand. When multiplied, these factors turn out to be equal to component Btu/total GDP, as shown in the following equation:

Contribution of component to total Btu/GDP

$$= \frac{\text{component Btu}}{\text{component \$}} \times \frac{\text{component \$}}{\text{total GDP}} = \frac{\text{component Btu}}{\text{total GDP}}$$

TABLE 7-3 Final-demand Mix Effect on Btu/Dollar Ratios for Specified Final-demand Components and Total Btu/GDP Ratios, 1970

GDP and final-demand components	U.S.	France	W. Germany	Italy	U.K.	Japan
			billion dollars GDP			
Consumption and fixed capital expenditures	831.6	180.0	202.1	127.6	157.7	286.8
Government and "other" components, total	152.2	19.5	33.0	12.4	24.8	51.6
Government compensation	72.6	8.5	11.3	6.7	15.9	30.9
All other (government commodities, net exports, net inventory change)	79.6	11.0	21.7	5.7	8.9	20.7
Total GDP	983.8	199.4	235.1	140.0	182.5	338.4
			percentage distribution of purchases			
Consumption and fixed capital expenditures	84.5	90.2	86.0	91.1	86.4	84.7
Government and "other" components, total	15.5	9.8	14.0	8.9	13.6	15.3
Government compensation	7.4	4.3	4.8	5.4	8.7	9.1
All other (government commodities, net exports, net inventory change)	8.1	5.5	9.2	3.5	4.9	6.1
Total GDP	100.0	100.0	100.0	100.0	100.0	100.0

	energy consumption in quadrillion Btus [derived]					
Consumption and fixed capital expenditures	59.8	10.1	11.8	6.8	9.7	15.9
Government and other components, total	4.6	0.6	1.2	0.3	0.5	1.2
Government compensation	0.0	0.0	0.0	0.0	0.0	0.0
All other (government commodities, net exports, net inventory change)	4.6	0.6	1.2	0.3	0.5	1.2
Total Btus (derived)	64.3	10.7	13.1	7.1	10.2	17.1
	ratio, thousand Btus per dollar					
Consumption and fixed capital expenditures	71.9	56.1	58.5	52.9	61.5	55.6
Government and other components, total	30.0	32.2	37.7	26.2	20.6	23.0
Government compensation	0.0	0.0	0.0	0.0	0.0	0.0
All other	57.2	57.2	57.2	57.2	57.2	57.2
Total Btu/GDP	65.4	53.8	55.5	50.6	56.0	50.6
	index of ratio, United States = 100					
Consumption and fixed capital expenditures	100.0	78.0	81.3	73.6	85.6	77.4
Government and other components	100.0	107.6	125.7	87.6	68.7	76.7
Total Btu/GDP	100.0	82.2	84.9	77.3	85.6	77.4
	percentage difference from United States					
Total Btu/GDP, reflecting difference in mix of all final-demand components	—	−17.8	−15.1	−22.7	−14.4	−22.6

Note: Blank = not applicable. Detail may not add to total due to rounding.

TABLE 7-4   Contribution of Components to Total Final-demand Mix Effect on Btu/GDP Ratios, Relative to United States, 1970
(percentage difference from United States)

Components	France	W. Germany	Italy	U.K.	Japan	Average of five countries
Consumption						
Total energy purchases	−25.6	−22.6	−26.6	−14.4	−32.4	−24.3
Fuel and power	−12.0	−10.1	−12.7	−2.8	−16.3	−10.8
Gasoline	−13.7	−12.5	−13.8	−11.6	−16.1	−13.5
Nonenergy purchases	1.0	−5.5	2.5	2.4	−0.4	0.0
Total consumption	−24.6	−28.1	−24.0	−12.0	−32.8	−24.3
Fixed capital expenditures						
Total fixed capital	9.1	12.1	4.9	0.4	12.0	7.7
Construction	7.4	10.1	5.7	0.3	9.4	6.6
Producer durables	1.7	2.0	−0.8	0.1	2.6	1.1
Total all other components	−2.3	1.0	−3.5	−2.8	−1.7	−1.9
Total Btu/GDP	−17.8	−15.1	−22.7	−14.5	−22.6	−18.5

Note: Detail may not add to total due to rounding.

The difference between the ratio for each component in other countries and the comparable ratio for United States represents that part of total final-demand mix effect on the aggregate Btu/GDP ratio, relative to U.S., attributable to each category.

This formulation was followed, using the relevant figures in tables 7-2 and 7-3, to develop estimates of the contributions of each final-demand component to the percentage difference in total Btu/GDP ratios relative to United States. The results, shown in table 7-4, have been largely foreshadowed by the previous discussions of the estimates (particularly in table 7-2) and by the inferences drawn regarding the role of the various components on the effect of final-demand mix on Btu/GDP ratios. The major finding is that consumer energy purchases are largely responsible for the lower aggregate Btu/GDP ratios in other countries due to final demand mix. On the average, these purchases account for a decline of about 24 percent in the Btu/GDP ratios of other countries, relative to that of United States—substantially more than the 18.5 percent net decline in the ratio due to all components. Within the consumer energy purchase component, gasoline purchases account for about 56 percent and fuel and power for about 44 percent. Part of the decline caused by consumer energy purchases served to offset the increase in the ratio attributable to fixed capital expenditures. On the average, the capital expenditures, particularly in the construction category, increased the overall ratios by almost 8 percent. The net effect of these two components (consumer energy purchases and fixed capital expenditures) accounts for most of the difference in aggregate Btu/GDP ratios in other countries, relative to United States. The effect of consumer nonenergy purchases is, on the average, negligible, but purchases by the group of components including government, net exports, and net inventory change serve to reduce the overall ratios by about 2 percent.

The effect of the various components on overall Btu/GDP ratios is fairly uniform among the countries in terms of direction and relative proportion, but not necessarily in the size of the effects. Specifically, almost all the countries show large reductions in Btu/GDP ratios due to consumer energy purchases and significant but substantially smaller increases due to fixed capital expenditures. The "all other" component group is also fairly uniform, with most of the countries (the exception being W. Germany) indicating reductions in the overall ratio attributable to this group. Nonenergy consumer purchases, however, show considerable diversity, with the figures for West Germany and Japan indicating reductions due to this component, the other three countries showing increases, and the average showing no effect.

The important role of direct energy purchases by households in reducing energy consumption per dollar of final demand in other countries, may seem surprising in view of the very small proportion of total purchases accounted for by direct energy consumption purchases. In the United States, consumer purchases of fuel and power, and gasoline amount to only 4.5 percent of the total GDP. In the other countries, they account for much lower proportions, ranging from .9 percent in Japan to 2.9 percent in the United Kingdom, with the average about 1.8 percent.

The explanation for this seeming disproportionate effect of a relatively small component of the total purchases on the aggregate Btu/dollar ratio is, as noted earlier, the extremely high energy-intensity ratio for the direct energy consumption items compared to the average for the other components of final demand. The average Btu/dollar ratio for all the nonenergy components of final demand is 40,900 Btus per dollar of purchases in the United States. In contrast, the average for the energy items is 581,000 Btus per dollar, about fourteen times as much. Compared to the enormous disparity between the average Btu/dollar ratio for the nonenergy components as a whole and that for the energy items, the difference among the nonenergy components is quite small, with the construction category having the highest ratio at 64,000 Btus per dollar, a little more than double the 30,000 Btu/dollar ratio for government purchases, net exports, net inventory change, the groups having the lowest ratio.

The reason for the much lower energy intensity ratios for the nonenergy components of final demand compared to the household energy purchases is that energy used to produce, transport, and distribute a final product represents only a small part of the total ultimate value of that product. The largest part consists of the value added by each industry (other than the energy industries) in the long chain from the raw material stage to the purchase of the final product. A great deal of value has to be added in addition to energy in order to make the final product. In contrast, the value of energy that is used directly by the final consumer for heating, lighting, and automobile transportation consists largely in the energy itself and not in the further production of nonenergy goods or services. The fact that the direct energy purchases have such a relatively high energy intensity ratio and that these items account for a much smaller proportion of final demand in other countries than they do in the United States is the major factor underlying the lower aggregate Btu/dollar ratios in other countries due to final-demand mix.

Proportion of Total Difference in Energy/GDP Ratios Accounted for by Final-demand Mix

Given the findings that differences in final-demand mix have the effect of reducing energy/GDP ratios in other countries by about 18.5 percent below the U.S. ratio (with the reductions ranging from about 15 to 23 percent among the various countries), the question remains as to how much of the *actual* difference in the ratios is explained by the final-demand mix effect and how much is left to be explained by other factors. The answer to this question can be seen in table 7-5, which compares the derived energy/GDP ratios reflecting only differences in final-demand mix with the actual ratios, with both of them converted to index form (U.S. = 100). The actual energy/GDP ratios for 1970 are based on estimates of total energy consumption and GDP for each country.[5] These are de-

---

[4] The estimates of energy consumption were converted from tons of oil equivalent (used in the OECD energy statistics) to British thermal units in order to be consistent with the energy measure used in the input–output analysis. The conversion factor used was 40 million Btus per ton of oil equivalent. Based on the OECD data, the derived estimate of energy consumption for United States in 1970 is quite close to the Bureau of Mines' estimate for that year, being only about 4 percent lower. The GDP estimates were derived from the data in the *ICP Report,* and were based on the GDP measure in which the quantities of all countries were weighted with U.S.

TABLE 7-5 *Proportion of Total Difference in Energy/GDP Ratios Accounted for by Final-demand Mix Versus Other Factors, 1970*

Ratios and differences	France	W. Germany	Italy	U.K.	Japan	Average of five countries
total energy/GDP ratios, U.S. = 100						
Actual ratio	47.9	63.6	53.7	70.6	50.5	57.3
Derived ratio, final-demand mix difference	82.2	84.9	77.3	85.6	77.4	81.5
percentage difference from U.S.						
Actual ratio	−52.1	−36.4	−46.3	−29.4	−49.5	−42.7
Differences due to:						
Final-demand mix	−17.8	−15.1	−22.7	−14.4	−22.6	−18.5
Other factors (residual)	−34.3	21.3	−23.6	−15.0	−26.9	−24.2
percentage distribution of difference						
Final-demand mix	34.2	41.5	49.0	49.0	45.7	43.3
Other factors (residual)	65.8	58.5	51.0	51.0	54.3	56.7

Note: Detail may not add to total due to rounding.

rived from the same sources as those used in chapter 2 to calculate the 1972 overall energy/GDP ratios and comparisons.

The figures in table 7-5 indicate that final-demand mix accounts for about 43 percent of the total difference between the average energy/GDP ratio of the other countries and that of the United States, with the remaining 57 percent due to other factors. Specifically, the difference in the actual ratios amounts to 43 percent, and the 18.5 percent due to differences in final-demand mix quite coincidentally turns out to be 43 percent of the total difference. The proportion of the total difference in energy/GDP ratios due to final-demand mix falls within a range of 42 to 49 percent among the various countries, excluding France. Of the proportion of the total difference due to final-demand mix, the average for the four countries other than France is 46 percent; France is 34 percent.

The contribution of the mix effect of each final-demand component to the actual change in energy/GDP ratios can be calculated by utilizing the results of the analysis in the previous section and the finding that, on the average, final-demand mix accounts for about 43 percent of the total difference in the energy/GDP ratio in other countries relative to United

price weights rather than Fisher ideal average price weights. The 1970 indexes of energy/GDP ratios for the countries being compared in this chapter are almost identical with those for 1972, when GDP for both years are based on U.S. price weights.

States. The table below shows the percentage of total differences in energy/GDP ratios attributable to differences in final-demand mix and the residual (all other factors).[6]

		Percentage
Consumer purchases		
	Direct energy	56.9
	All other	0.0
Fixed investment		−18.0
Government and all other		4.4
Total mix effect		43.3
Residual (all other factors)		56.7
	Total difference	100.0

Of the total difference in Btu/GDP ratio in other countries due to all factors, the differences due to the mix effect of consumer direct purchases of energy account for 57 percent of the total, offsetting the increases in the ratios due to fixed capital expenditures (18 percent). Consumers' nonenergy purchases play a negligible role, and purchases by government and "all other" components account for about 4 percent of the total difference. The major role of consumer direct energy purchases in accounting for a substantial part of the total difference in energy/GDP ratios, relative to United States, is thus further underscored.

One final comment should be made regarding the residual difference in energy/GDP ratios relative to United States. This residual difference encompasses all the factors not included in the analysis of the effect of differences in final-demand mix on the overall ratios. Some of these factors have already been mentioned; they include differences in: (1) input–output coefficients among countries (including imports as proportions of total supply of "industry" output), (2) product mix within the 89 product groups used to analyze effect of final-demand mix on energy/GDP ratios, (3) patterns of transportation and distribution required to deliver final products from producers to purchasers, and (4) factors used to convert energy industry output requirements from dollars to Btus. In addition, all these factors interact with each other and with the differences in the overall energy/GDP ratios attributable to final-demand mix. The interaction effect could be significant and, if allocated on some basis to each of the factors, modify the proportion of the total accounted for by differences in final-demand mix.

[6] A minus sign in the context of this text table indicates that the component contributed to an increase in the ratios relative to United Sates.

Summary

Energy/GDP ratios in other countries may differ from that in the United States because the mix of final goods and services produced (GDP) by these countries may be different from that in the United States, with a higher or lower proportion consisting of energy-intensive products *or* the mix may be about the same but the energy required to produce, transport, and distribute the individual products may be higher or lower than in the United States. Input–output analysis was used to measure the extent to which the differences in final-demand mix of consumer and fixed investment goods and services for France, Germany, Italy, United Kingdom, and Japan contribute to the total difference in energy/GDP ratios in these countries relative to that in the United States. A more aggregate approach was used for the remaining components of GDP, including government purchases, net exports, and net inventory change.

The energy/GDP ratios for the countries included in the analysis are, on the average, about 43 percent below the level of the U.S. ratio (in 1970). The range among the countries was quite wide, with the United Kingdom closest at 29 percent below the U.S. ratio and with France furthest away at 52 percent below that of the United States. Of the 43 percent difference in the average level of the energy/GDP ratios relative to United States, about 18.5 percentage points or 43 percent (coincidentally) of the total difference was attributable to the differences in mix of final-demand components in other countries relative to United States.

The final-demand mix effect largely represents the net effect of two partially offsetting differences in the components of final demand. The difference in fixed investment mix would have increased Btu/GDP ratios in other countries by about 8 percent; this, however, was more than offset by the effect of lower direct-energy purchases by consumers in other countries. The effect of this latter factor was to reduce the overall energy/GDP ratios by about 24 percent; the net effect was a reduction of about 17 percent in overall energy/GDP ratios in the other countries relative to the United States. Government and "all other" purchases accounted for another 2 percent of the total 18.5 percent difference due to final-demand mix.

The remaining percentage, or 57 percent of the total difference, was due to a number of factors, primarily (1) differences in the direct and indirect energy required per dollar of end product to produce, transport, and distribute individual categories of final goods and services relative to the United States and (2) interaction among all the factors.

CHAPTER 8

Consumption Expenditures for Energy and Related Items

ONE OF THE KEY FINDINGS of the analysis in the previous chapter on the effect of differences among countries in final-demand mix on disparities in energy/GDP ratios is the large proportion of the total mix effect accounted for by the differences in direct-energy purchases by consumers for household operations and personal automobile transportation. The objective of the analysis in this chapter is to provide additional insight into some of the factors that help explain the much lower proportion of total consumer expenditures devoted to direct purchases of energy in other countries relative to the United States. This chapter analyzes such purchases within the context of total consumption expenditures, based on the *ICP Report*.[1] This report gives detailed information on the relative quantities and prices of direct consumer expenditures for energy, as well as on their purchases of energy-related items, such as household appliances, automobiles, automotive operating costs, and purchased transportation (for example, train and bus transportation), relative to similar purchases in the United States. Another advantage of using the *ICP Report* is that it provides this information within the same national accounts framework and purchasing-power-parity approach that underlies the estimates of GDP used in chapter 7. In this connection, it should be noted that the estimates in the *ICP Report* refer primarily to 1970, as does the discussion in this chapter. This is in contrast to most of the analyses of energy/GDP use patterns in the present study, which are based on estimates for 1972.

---

[1] Irving B. Kravis, Zoltan Kenessey, Alan Heston, and Robert Summers, *A System of International Comparisons of Gross Product and Purchasing Power* (Baltimore, published for the World Bank by The Johns Hopkins University Press, 1975).

TABLE 8-1   Measures of Per Capita Consumption, 1970

Currencies	Based on system of nat'l. accounts (SNA)[a]	Based on ICP study	Index (SNA = 100.0)
U.S. ($)	3,019.73	3,271.73	108.3
France (FR)	9,497.08	10,302.14	108.5
W. Germany (DM)	6,641.94	6,050.57	109.8
Italy (L)	683,983.	741,489.	108.4
U.K. (£)	552.619	634.348	114.8
Japan (Y)	345,106.	367,718.	106.6

Source: ICP Report
[a] United Nations System of National Accounts.

Energy Purchases as a Percentage of Total Consumption Expenditures

How important are direct energy purchases in the mix of consumer goods and services in the United States and other industrialized countries included in the ICP study?[2] The answer to this question depends in part on the definition of consumption used in the ICP study. In that study, the relative importance of energy purchases is affected by an adjustment to total consumer expenditures that assures comparability among countries in expenditures for certain services provided by government, particularly health and education. Specifically: the way in which health and educational services are provided varies considerably among countries, with the proportion provided directly by government or by the private sector being the key factor involved. To assure that this factor would not distort the comparison of consumer expenditures among countries, the ICP study added all government expenditures for health and education to personal consumption expenditures to obtain a more comprehensive measure of expenditures for personal use, although these expenditures are not necessarily paid for directly by persons.[3]

Table 8-1 shows the estimates of personal consumption expenditures for 1970 as they are (1) normally shown, consistent with the UN System of National Accounts, (SNA) and (2) as they are after the adjustment in the ICP study for inclusion of government expenditures for health, education, and the like. Note that the adjusted measures are higher than the official estimates by varying percentages, with United States, France, and Italy about 8.4 percent higher; Germany somewhat higher at 9.8

[2] France, West Germany, Italy, United Kingdom, and Japan.

[3] Government recreational services and housing are also included in the ICP consumption category.

percent; but United Kingdom with its large government-supported medical service, at 14.8 percent higher; and Japan, much lower than the others, at 6.6 percent higher.

In this book, to assure consistency in the treatment of energy items within the context of the ICP study, the analysis of energy purchases in relation to total consumption expenditures refers to the ICP-adjusted measure of consumption expenditures, rather than the SNA measure. However, it will also be referred to as "personal consumption expenditures" (PCE), since by far the largest component consists of purchases by households.

The ICP study provides estimates of consumption expenditures per capita, both for broad categories of goods and services, and for more detailed items underlying the broad categories. From the viewpoint of identifying and analyzing the direct energy component of consumption expenditures, there are two categories that cover the energy purchases. The first is the expenditure for fuel and power for household operations. The second is for gasoline, oil, and grease, primarily used for automobile transportation (including gasoline for motor scooters and motorcycles).

Included in the fuel and power group of purchases and shown separately in the more detailed tables are electricity, gas, liquid fuels, and other fuels and ice. "Other fuels and ice" includes solid fuels (for example, coal). Unfortunately, separate information is not provided on ice so that it cannot be excluded from this category.

In addition to the various categories of energy purchases for personal use, there are a number that cover purchases of energy-related items such as appliances, automobiles, purchased transportation, and so forth. The estimates for the energy-related items are quite useful as they provide part of the explanation for different quantities of energy consumed by other industrialized nations, relative to the United States. At this point, however, the analysis will focus on the direct energy purchases, leaving the energy-related items for later discussion.

Table 8-2 is based on the ICP estimates of direct-energy purchases per capita in own-country currencies and prices, both for major categories and specific energy types. Estimates for other countries cannot be compared directly with the U.S. figures because they are stated in each country's own currency. They can, however, be related to each country's own per capita personal consumption expenditures to determine (1) whether the share of the consumer budget devoted to direct-energy purchases is about the same, higher, or lower than in the United States; and (2) whether any differences in energy/consumption ratios are general

TABLE 8-2 *Energy/Consumption Ratios: Energy Purchases in Relation to Total Consumption Expenditures, in Own-Country Quantities and Prices, 1970*

Ratios	U.S.	France	W. Germany	Italy	U.K.	Four Western European countries (unweighted average)	Japan
E/C ratio (energy purchases [E] ÷ total consumption [C])							
Panel A. Energy/consumption ratios							
Total energy purchases	6.7	5.0	6.3	5.4	6.4	5.8	3.2
Fuel and power	3.2	2.9	3.5	2.9	4.2	3.4	2.5
Electricity	1.5	1.0	1.5	1.3	1.9	1.4	1.2
Fuel	1.7	2.0	2.0	1.6	2.3	2.0	1.4
Gasoline, oil, grease	3.4	2.0	2.8	2.4	2.2	2.4	0.6
E/C ratio							
Panel B. Energy/consumption ratios (index, U.S. = 100)							
Total energy purchases	100.0	74.3	95.2	80.8	96.4	86.7	47.5
Fuel and power	100.0	90.6	108.8	91.2	130.3	105.2	78.7
Electricity	100.0	64.8	101.0	89.2	125.5	95.1	79.1
Fuel	100.0	112.5	115.4	93.0	134.3	113.8	78.4
Gasoline, oil, grease	100.0	59.2	82.5	71.0	64.8	69.4	18.3
Energy purchases							
Panel C. Percentage distribution of energy purchases							
Total energy purchases	100.0	100.0	100.0	100.0	100.0	100.0	100.0
Fuel and power	48.3	58.8	55.2	54.5	65.3	58.4	80.1
Gasoline, oil, grease	51.7	41.2	44.8	45.5	34.7	41.6	19.9
Total fuel and power	100.0	100.0	100.0	100.0	100.0	100.0	100.0
Electricity	46.0	32.9	42.7	45.0	44.3	41.2	46.2
Fuel	54.0	67.1	57.3	55.0	55.7	58.8	53.8

or largely confined to one or two specific energy types or countries. The energy/consumption ratios[4] are shown in panel A of table 8-2 and then compared in panel B to the U.S. ratios in the form of indexes, with the United States equaling 100.

Turning again to panel A of the table showing the percentage of total consumption accounted for by direct energy purchases, we find that in all the other countries a smaller proportion of the consumer budget is spent for energy than in the United States. The range is quite wide, from 3.2 percent for Japan to 6.4 percent for the United Kingdom (compared to 6.7 percent for the United States). However, the lower ratio for other countries consists of quite different patterns for the two major categories. Purchases of fuel and power for household operations in the other industrialized nations are, with the exception of Japan, about the same proportion of total consumption or even higher than in the United States. The ratio for the United States is 3.2 percent. France and Italy are slightly lower at 2.9 percent, and Germany, slightly higher at 3.5 percent. But the United Kingdom is substantially higher—4.2 percent. The unweighted average for the four Western European countries is 3.4 percent, slightly higher than the 3.2 percent for the United States. The ratio for Japan, at 2.5 percent, is considerably lower than the U.S. proportion of the consumer budget being spent for this purpose.

In contrast to the expenditure pattern for fuel and power—about the same as or higher than the United States—the proportion of total consumption being spent for gasoline, and the like, is substantially lower in all the other countries relative to the United States. The percentage for the United States is 3.4 percent. That for the other countries, excluding Japan, ranges from about 2.0 to 2.8 percent, with 2.4 the unweighted average. Japan spends a much smaller proportion of the consumer budget for gasoline, only 0.6 percent, implying a much heavier reliance in Japan on public transportation than in the United States and the other industrialized nations.

Looking at gasoline purchases another way, the United States is the only one of the six countries that devotes a somewhat higher proportion of the consumer budget for gasoline (3.4 percent) than for fuel and power (3.2 percent) for household operations. All the other countries allocate a smaller portion of total consumption expenditures for gasoline than for fuel and power—and generally by substantial margins.

The various energy/consumption ratios for each country relative to the United States, which are implicit in panel A, are made explicit

---

[4] Since population appears in both the numerator and denominator of the ratio, population cancels out.

in panel B in the form of indexes, with United States equaling 100. For the four Western European countries, total energy purchases as a proportion of total consumption expenditures is on the average about 15 percent less than that of the United States, with the fuel and power proportion about 5 percent higher and the gasoline ratio 30 percent lower. Japan is substantially lower for both major categories of energy purchases —more than 15 percent less than the United States for fuel and power, and more than 80 percent less for gasoline.

The relative importance of energy use for household operations as contrasted with use for personal automobile transportation is seen even more clearly in panel C of table 8-2, which provides a percentage distribution of total energy use between the two major components. For the United States, the distribution is almost even, with fuel and power accounting for slightly less than half, and gasoline for a little more than half. The distribution for the four Western European countries shows a substantially higher proportion being devoted to fuel and power, almost 60 percent, and a correspondingly lower percentage for gasoline, closer to 40 percent. Japan, because of its relatively lower use of gasoline, is very different: only 20 percent for gasoline and 80 percent for fuel and power.

Regarding the components of the fuel and power category, panel C in table 8-2, we find that with the exception of France, electricity and fuels account for about the same proportion of the total fuel and power category in all the countries—about 45 percent for electricity and 55 percent for fuels. Within the fuel component, the relative importance of the specific types of fuels (not shown separately in table 8-2) varies from country to country, but all the other countries spend proportionately more for solid fuel (coal and briquettes) than does the United States and, offsetting this, somewhat less for gas and liquid fuel.

Conversion of Energy/Consumption Ratios To U.S. Prices

In the previous section, it was found that as a proportion of total consumption expenditures, energy consumption for household purposes (in terms of each country's prices) was, with the exception of Japan, about the same as or higher than in the United States. However, the proportion of the consumer budget being spent for gasoline, oil, and grease was uniformly lower by fairly large margins.

These estimates reflect price as well as quantity differences and therefore need to be stated in comparable prices in order to shed any light on the factors underlying the differences in aggregate energy/GDP ratios. The estimates of per capita energy purchases and per capita consumption expenditures have therefore been converted to constant prices, based on the detailed information given in the *ICP Report*. The measures for the individual categories may vary depending on the weighting system used. In contrast to the weighting system underlying the GDP measure described in chapter 2 to derive aggregate energy/GDP ratios—which was based on the average price weights of the United States and each country in turn (the "ideal" measure)—the quantity comparisons developed in this chapter are based on U.S. price weights. The reason is to assure consistency with the analysis in the previous chapter, which is based on the use of a U.S. input–output table for 1970. Consistency with this table requires that final-demand purchases, including purchases of energy items, be stated in U.S. prices.

The estimates of per capita energy purchases and per capita consumption expenditures (in U.S. prices) and the ratio of the two are shown in table 8-3. The estimates indicate much lower per capita energy purchases by other countries relative to United States, but there is considerable variation among the countries. Japan shows the lowest level, at 13 percent of U.S. per capita energy purchases. The average for the four European countries is 31 percent, with Italy at the low end of a range, at only 18 percent, and United Kingdom at the upper end, at 43 percent.

Within the total per capita energy purchases, there is a substantial difference between the two major categories, fuel and power; and gasoline, oil, and grease. Relative to United States, the quantities of both are lower, but by far the largest difference is in gasoline. The average per capita purchase by the four European countries for fuel and power is about 45 percent of the U.S. level, whereas gasoline is only 17 percent. In the case of Japan, the difference is even more striking, 23 percent for fuel and power relative to United States and only 3.5 percent for gasoline.

Part of the explanation for the substantially lower levels of per capita energy purchases may be in the fact that all of the other countries in this comparison have lower levels of total consumer expenditures per capita, and even if energy use were to account, in real terms, for the same proportion of the consumer budget as in the United States, the absolute levels of per capita energy purchases would still be lower than in United States. (The average per capita consumption expenditure for the four European countries is only about two-thirds of the U.S. level; with a range of 55 percent for Italy and 77 percent for France. Japan is close to the lower end of the range for the other four countries—at 56 percent.) In order to adjust for this factor, the per capita energy purchases have been divided by the per capita total consumption expenditures, both in U.S. prices. The resulting energy/total consumption ratios are shown in panel B of table 8-3 and as indexes (U.S. = 100) in panel C. These ratios, based on expenditures in U.S. prices, can then be compared to the same ratios in own-country prices. The latter appear in table 8-2.

Adjusting for lower levels of per capita total consumption expenditures results in energy/consumption ratios which are closer to the U.S. ratio than the initial comparison of per capita energy purchases, but still substantially below the proportion of the U.S. consumer budget devoted to energy purchases. The average for the four European countries is about 45 percent of the U.S. level. (Japan is even lower at 23 percent.) This is a sharp drop from the 87 percent average level of the energy/consumption ratios for the four countries, relative to United States, when stated in own-country prices, implying much higher prices for energy items in the other countries relative to those in United States.

When the comparison is based on U.S. prices rather than on own-country prices, the reduction in the energy/consumption ratio holds for the major components of energy purchases. It will be recalled that in terms of each country's own prices, the ratios for fuel and power in other countries, with the exception of Japan, were about the same or higher than in United States. Based on purchases in U.S. prices, the average ratio for the four European countries drops to about two-thirds of the U.S. level. Similarly, the average energy/consumption ratio for gasoline purchases in table 8-2 was almost 70 percent of the U.S. level. In U.S. prices, it drops to only 25 percent. For Japan, the decline is from 18 percent of the U.S. level to only 3. 5 percent.

The much lower energy/consumption ratios for gasoline relative to United States than for fuel and power suggest rather strongly that an important part of the explanation for lower energy consumption in other countries relative to total GDP, lies in the much smaller share of their purchases going for gasoline. (As we shall see later, this lower dependence on the automobile in other countries is reflected in the larger proportion of the consumer budget allocated to the use of public transportation.)

The smaller share of gasoline in total energy purchases in other countries relative to United States is shown most clearly in panel D in table 8-3. As compared with the 52 percent share in United States, the average for the four European countries is only 29 percent, for Japan only 14 percent.

TABLE 8-3 Per Capita Energy Purchases in Relation to Total Per Capita Consumption Expenditures, in Own-Country Quantities and U.S. Prices, 1970

Item	U.S.	France	W. Germany	Italy	U.K.	Four Western European countries (unweighted average)	Japan
				Panel A			
Total per capita consumption							
Dollars	3,271.73	2,509.42	2,208.42	1,799.45	2,290.21	2,201.88	1,845.26
Index, U.S. = 100	100.0	76.7	67.5	55.0	70.0	67.3	56.4
Per capita energy purchases							
Total (dollars)	217.85	60.20	76.67	38.39	93.75	67.25	28.56
Fuel and power	105.21	42.40	53.13	27.35	69.65	48.13	24.62
Electricity	48.39	12.87	23.66	13.02	35.66	21.30	13.16
Fuel	56.82	29.49	29.41	14.36	33.99	26.81	11.41
Gasoline, oil, grease	112.64	17.80	23.54	11.04	24.10	19.12	3.94
Total (index, U.S. = 100)	100.0	27.6	35.2	17.6	43.0	30.9	13.1
Fuel and power	100.0	40.3	50.5	26.0	66.2	45.7	23.4
Electricity	100.0	26.6	48.9	26.9	73.7	44.0	27.2
Fuel	100.0	51.9	51.8	25.3	59.8	47.2	20.1
Gasoline, oil, grease	100.0	15.8	20.9	9.8	21.4	17.0	3.5

				Panel B			
Energy/consumption ratios							
Total (percentage)	6.7	2.4	3.5	2.1	4.1	3.0	1.5
Fuel and power	3.2	1.7	2.4	1.5	3.0	2.2	1.3
Electricity	1.5	0.5	1.1	0.7	1.6	1.0	0.7
Fuel	1.7	1.2	1.3	0.8	1.5	1.2	0.6
Gasoline, oil, grease	3.4	0.7	1.1	0.6	1.1	0.9	0.2
				Panel C			
Total (index, U.S. = 100)	100.0	36.0	52.1	32.0	61.5	44.8	23.2
Fuel and power	100.0	52.5	74.8	47.3	94.6	67.3	41.5
Electricity	100.0	34.7	72.4	48.9	105.3	65.3	48.2
Fuel	100.0	67.6	76.7	45.9	85.4	68.9	35.6
Gasoline, oil, grease	100.0	20.6	31.0	17.8	30.6	25.0	6.2
				Panel D			
Percentage distribution of energy purchases							
Total energy	100.0	100.0	100.0	100.0	100.0	100.0	100.0
Fuel and power	48.3	70.4	69.3	71.2	74.3	71.3	86.2
Gasoline, oil, grease	51.7	29.6	30.7	28.8	25.7	28.7	13.8
				Panel E			
Total fuel and power	100.0	100.0	100.0	100.0	100.0	100.0	100.0
Electricity	46.0	30.4	44.5	47.6	51.2	44.3	53.5
Fuel	54.0	69.6	55.4	52.5	48.8	55.7	46.3

Source: ICP Report.

With the exception of France, within the fuel and power category (panel E), the distribution between electricity and fuel is not too different from the 46–54 percent split in the United States. France uses relatively more fuel and substantially less electricity than the other countries. Finally, analysis of the composition of fuel purchases among the three specific types, that is, gas, liquid fuel, and solid fuel, indicates that the earlier finding that consumers in other countries use proportionately more solid fuel such as coal and briquettes than in United States still holds, even when the estimates are converted to U.S. prices. Use of gas and liquid fuels is proportionately less than in United States. No one of the three types of fuel used for household operations is more important than any other across all countries. This suggests that fuel use by households ought to be looked at in the aggregate and that the particular composition of that aggregate will vary considerably from country to country depending on domestic energy resources, relative prices, costs of imports, and stock of domestic heating and cooking equipment that use different types of fuels.

Comparisons of Relative Prices and Constant Dollar Purchases of Energy

In the previous section it was suggested that one of the major reasons for the much lower purchases, in U.S. constant price terms, of energy by consumers in other industrialized countries relative to the United States were the higher relative prices for energy in these countries. This is implicit in the fact that the relative purchases of energy items, in U.S. constant prices, are uniformly much lower than the same purchases stated in own-country prices. In this section, these relationships will be made explicit.

As previously indicated the ICP study provides deflators (purchasing power parities) both for broad categories and detailed items of personal consumption expenditures, for converting these purchases into quantity relatives, with United States equaling 100. The initial estimates of purchases are given in each country's own currency and price level. The deflators therefore reflect both differences in currency exchange rates and price differences. In order to analyze the role of relative prices in explaining part of the lower purchases of energy, in real terms, by other countries relative to the United States, it is necessary to separate the price differences from the currency differences. For this purpose, the overall purchasing power ratio for all consumer goods and services from the ICP study was used as the best approximation of the internal "exchange rate" of currencies between United States and other countries.

Using the aggregate consumption purchasing-power-parity ratios, we are able to derive the relative price changes for broad categories and individual items of energy purchases. These can then be compared to the relative energy/consumption quantity ratios already given in table 8-3, except that additional detail will be provided by specific types of fuels.

Relative price for this purpose is defined as the price of a specific item in country X relative to the price of that item in United States, divided by the average price of all consumer goods and services in country X relative to the price of all consumer items in United States. In an analogous way, relative quantity is defined as the quantity of an item purchased in country X relative to the quantity of that item purchased in United States, divided by the aggregate quantity of all consumer goods and services purchased in country X relative to the quantity of consumer goods and services purchased in United States.

As used in this chapter the measures of relative prices and quantities are "relative" in a double sense; relative to the prices and quantities of the same items in United States and also relative to the level of prices and quantities of total consumption purchases in each country compared to United States. They are, therefore, different from the conventional measures of relative prices and quantities, which are limited to the price and quantity differences for the specific items without regard to whether these differences are more or less than that for all consumer items.

The comparison of relative prices and quantities is shown in table 8-4. In panels A and B are indexes of relative prices and quantities, with U.S. = 100. Panels C and D convert these indexes into percentage differences from the U.S. level.[5]

The results can be summarized quite briefly. With one minor exception, prices of energy in the other countries covered in the ICP study are higher than in the United States and usually by substantial margins. Prices of energy used as fuel and power for household operations range from 38 to 93 percent higher than in United States. The average for the four

---

[5] It should be added as a technical note that consistent with the usual index number factoring of a value index into two components that are multiplicative, one of the factors takes the form of a Laspeyres index with base-year or base-country weights. The other is a Paasche index with other-year or other-country weights. In this instance, the quantity relatives are based on constant U.S. price weights and are therefore of the Laspeyres type; the price relatives are based on the other-country quantity weights, which change with each binary comparison of other countries and United States, and are therefore Paasche type indexes.

TABLE 8-4 Prices and Quantities of Energy Purchases, Relative to United States, 1970

Relative prices and quantities	France	W. Germany	Italy	U.K.	Four Western European countries (unweighted average)	Japan
			Panel A. index, United States = 100			
Relative prices						
Fuel and power	172.3	145.5	193.0	137.5	162.0	189.9
Electricity	186.6	139.2	182.5	119.1	156.9	163.8
Fuel	166.4	150.5	202.6	157.3	169.2	220.2
Gas	225.3	251.2	297.1	321.3	273.7	681.9
Liquid	136.5	111.0	156.6	159.9	141.0	207.5
Solid	144.3	148.5	161.4	103.6	139.5	94.5
Gasoline, oil, grease	287.3	266.4	398.3	211.9	291.0	297.0
			Panel B. percentage			
Relative quantities (E/C ratios)						
Fuel and power	52.5	74.8	47.3	94.6	67.3	41.5
Electricity	34.7	72.4	48.9	105.3	65.3	48.2
Fuel	67.6	76.7	45.9	85.4	68.9	35.6
Gas	39.5	25.8	28.4	37.4	33.0	12.2
Liquid	57.8	80.9	55.5	15.6	52.5	16.7
Solid	305.6	392.6	113.8	764.7	394.2	288.3
Gasoline, oil, grease	20.6	31.0	17.8	30.6	25.0	6.2

Panel C. percentage difference from United States

Relative prices

Fuel and power	72.3	45.5	93.0	37.5	62.1	89.9
Electricity	86.6	39.2	82.5	19.1	56.9	63.8
Fuel	66.4	50.5	102.6	57.3	69.2	120.2
Gas	125.3	151.2	197.1	221.3	173.7	581.9
Liquid	36.5	11.0	56.6	59.9	41.0	107.5
Solid	44.3	48.5	61.4	3.6	39.5	-5.5
Gasoline, oil, grease	187.3	166.4	298.3	111.9	191.0	197.0

Panel D. percentage difference from United States

Relative quantities (E/C ratios)

Fuel and power	-47.5	-25.2	-52.7	-5.4	-32.7	-58.5
Electricity	-65.3	-27.6	-51.1	5.3	-34.7	-51.8
Fuel	-32.4	-23.3	-54.1	-14.6	-31.1	-64.4
Gas	-40.5	-74.2	-71.6	-62.6	-67.0	-87.8
Liquid	-42.2	-19.1	-44.5	-84.4	-47.5	-83.3
Solid	205.6	292.6	13.8	664.7	294.2	188.3
Gasoline, oil, grease	-79.4	-69.0	-82.2	-69.4	-75.0	-93.8

European countries is 62 percent above that for United States. Within the fuel and power category, gas is by far the most expensive type of energy for all the countries, with the ranking of the relative prices of electricity, and liquid and solid fuels varying from country to country. For the four European countries as a whole, the average prices of the four types of energy in the fuel and power category are higher than in the United States by 174 percent for gas; 57 percent for electricity; and liquid and solid fuels about the same, 40 percent.

The prices of gasoline, oil, and grease in the other countries are on the average about 200 percent higher than in the United States, with United Kingdom at the lower end of the range (112 percent higher) and Italy at the upper end (about 300 percent higher).

With energy prices so much higher than in the United States, one would expect to find a substantially lower proportion of consumer income being allocated to energy purchases, in quantity terms. Indeed this is exactly what happens. As previously noted in the discussion focusing on table 8-3, all of the countries show much smaller purchases of energy as a proportion of total consumption expenditures than does the United States, with the exception of purchases of solid fuel, like coal. Although the prices of solid fuel are either higher or about the same as in United States, the relative quantity of foreign purchases of solid fuel relative to total PCE in constant prices is much higher. The explanation for this phenomenon is not interfuel substitution due to relative fuel prices within each country (because solid fuel is not uniformly that much cheaper than liquid fuels, for example), but to the energy-related investment in the home, which may be more important than the direct fuel cost. These energy-related investments may have been built up over an extended period of time incorporating the types of fuel and power used for such activities as heating, cooking, and lighting. These investments are not easily changed. This involves questions of relative cost of energy-related categories such as costs of appliances in other countries relative to United States. Such questions will be covered in the next section.

In addition, household use of solid fuel has been almost entirely superseded in the United States for a variety of reasons related to convenience and cleanliness, so that price in this case clearly is not the major consideration. The other countries may be following U.S. practice in abandoning coal and briquettes for household use, but they still have a little way to go.

Aside from this type of fuel, all other energy items show substantially lower energy/consumption ratios than the United States, and, in general, the higher the price, the lower the ratio.

In the case of gasoline, the relative price is usually much higher than the relative price of household energy-using items and, correspondingly, the energy/consumption ratios relative to United States are much lower than the corresponding ratios for household energy use.

Of course, the price of gasoline is only one factor affecting the quantities purchased. The relative cost of automobiles, motorcycles, and the like, and the cost of motor vehicle operation are additional factors affecting decisions to use private personal transportation. Moreover, the availability and cost of public transportation also affect consumer decisions about whether to favor personal transportation involving the direct use of energy over public transportation, which represents a purchase of services from the transportation sector. These are discussed in the next section and in chapter 5.

Relative Prices and Energy/Consumption Ratios of Energy-related Items

There are many factors which affect the quantity of energy used by consumers in addition to relative prices. Clearly one of the more important considerations is the cost of the heating system, appliance, or automobile, that affects the kind and number of consumer durables purchased and, indirectly, the amount of energy used to operate the consumer durables. In addition to the detailed data on energy purchases, the ICP study also provides information on relative prices and quantities purchased for a number of these energy-related items. As in the case of energy purchases, we can group these energy-related items into two categories; those related to household operation and those related to personal transportation.

In the first group are various types of appliances such as refrigerators, and washing, cooking, heating, and cleaning appliances. In the second group, those related to personal transportation, are three categories of items. First are purchases of motor cars and other vehicles for personal transportation. Second are operating costs of such vehicles (including tires, tubes and accessories, repair charges, parking, tolls, and the like, as well as gasoline, oil, and grease). Third are the various forms of public transportation which are alternatives to personal vehicle transportation and which are less energy intensive per passenger-mile than personal vehicles.

With one exception all of these categories are shown as separate items in the *ICP Report,* with their associated price and quantity relatives. The exception is appliances data, which are shown separately for six different types of appliances. For the purpose of this report, however, estimates of price and quantity relatives have been developed for the appliance group as a whole.

The pattern which emerges from the figures in table 8-5 follows that of relative price and energy/consumption ratios for direct purchases of energy. The price of energy-related purchases such as appliances, autos, and operating costs of motor vehicles are much higher in other countries than in United States, and the relative quantities of such purchases, as a share of total consumption expenditures, are substantially lower.

Specifically, the average prices for these items in the four European countries is about 46 percent higher than in United States for appliances, 36 percent for autos, and 46 percent for auto operating expenses. The relative share of the consumer budget allocated to such purchases is 33 percent less than in United States for appliances, 60 percent less for autos, and 44 percent less for auto operating costs.

Relative prices in Japan for these same items are even higher, about double in the case of appliances, 80 percent higher for autos, and 125 percent for auto operating costs. The relative purchases of such items were correspondingly lower, particularly for automobile and automobile-related costs.

Quite consistent with this pattern is the finding that the price of public (purchased) transportation is lower in other countries relative to United States, particularly in Japan where it is 63 percent lower and the relative quantity/PCE purchases are much higher. The use of public transportation rather than private transportation would tend to reduce the overall energy/GDP ratio because public transportation is less energy intensive.

Summary

To summarize, the five countries included in this analysis devote a substantially smaller proportion of the consumer budget (in constant U.S. prices), to purchases of energy items for household operations and personal auto transportation than does the United States. Purchases of fuel and power for household operations account for 3.2 percent of total consumption expenditures in United States, but average 2.2 percent for the four Western European countries and only 1.3 percent for Japan. The disparity for gasoline is even greater. The U.S. proportion is 3.4 percent, the four Western European countries average slightly less than 1 percent, and Japan a miniscule two-tenths of one percent.

TABLE 8-5 *Purchases of Energy-related Items: Prices and Quantities Relative to United States, 1970*

Relative prices and quantities	France	W. Germany	Italy	U.K.	Four Western European countries (unweighted average)	Japan
			index, United States = 100			
Relative prices						
Appliances	160.6	132.1	128.4	162.1	145.8	198.6
Autos, etc.	127.0	123.6	134.7	156.7	135.5	180.4
Auto operating costs	150.9	171.4	143.4	118.4	146.0	224.6
Purchased transportation	93.2	87.4	71.4	96.0	87.0	37.2
Relative quantities[a]						
Appliances	72.3	87.9	49.7	59.4	67.3	58.2
Autos, etc.	41.9	46.4	37.5	37.7	40.8	10.3
Auto operating costs	57.8	51.7	48.2	65.7	55.8	9.2
Purchased transportation	179.4	228.6	272.9	312.6	248.3	1,131.0
			percentage difference from United States			
Relative prices						
Appliances	60.6	32.1	28.4	62.1	45.8	98.6
Autos, etc.	27.0	23.6	34.7	56.7	35.5	80.4
Auto operating costs	50.9	71.4	43.4	18.4	46.0	124.6
Purchased transportation	−6.8	−12.6	−28.6	−4.0	−13.0	−62.8
Relative quantities[a]						
Appliances	−27.7	−12.1	−50.3	−40.6	−32.7	−41.8
Autos, etc.	−58.1	−53.6	−62.5	−62.3	−59.2	−89.7
Auto operating costs	−42.2	−48.3	−51.8	−34.3	−44.2	−90.8
Purchased transportation	79.4	128.6	172.9	212.6	148.3	1,031.0

Source: ICP Report.

[a] The ratios of constant dollar purchases of energy-related items as a percentage of total consumption expenditures relative to similar ratios in United States.

Part of the explanation for the relatively lower use of energy by consumers in the other countries may be inferred from the fact that consumers' expenditures for energy-using consumer durables in other countries are also much lower than in United States, again as a proportion of the consumer budget. For the four Western European countries, purchases of household appliances in 1970, as a proportion of total consumption, averaged only about two-thirds of the U.S. percentage, and less than 60 percent for Japan. The disparity for autos is much greater, an average of 40 percent of the U.S. ratio for the European countries included in this comparison, and only 10 percent for Japan. On the other hand, expenditures for purchased transportation, which is less energy intensive than automobile transportation, are proportionately much higher in other countries—by a wide margin—than in the United States.

Underlying the lower proportion of purchases of energy and related energy-using consumer durables, is a general pattern of substantially higher relative prices for both energy items and consumer durables such as appliances and autos. In addition, higher prices for automobile operating costs and lower prices for purchased public transportation in other countries provide additional incentives to use less-energy-intensive modes of transportation than the automobile.

How much of the lower energy purchases in other countries as a proportion of total consumer expenditures, relative to that of United States, can be attributed to higher relative prices, both for energy and energy-related items, is difficult to determine. The almost uniform and strong inverse relationship between relative prices and relative quantities of energy purchased suggests that prices do play a major role in lower energy use by consumers in other countries relative to United States. However, the limited sample of countries covered makes it difficult to develop a statistically significant measure of this relationship. Some of the considerations involved in assessing the extent to which relative prices, income, and related factors contribute to the lower quantities of energy used relative to United States are discussed further in chapter 9.

PART IV

Synthesis

Findings and Interpretation

DATA FROM VARIOUS nations make it clear that variations in the per capita use of energy are not tightly linked to differences in economic development measured by output per capita. Hence, our principal analytical effort has been to explore the extent to which these variations are rooted in differences in the composition of national output, on the one hand, or differences in the energy intensity associated with given economic processes or activities, on the other. Beyond identification of these respective contributory elements in a purely statistical sense, some attempt was made to probe—at least in a selective fashion—the underlying economic factors at work. A two-pronged approach sought to deal with the topic both by examining specific economic sectors of the countries (chapters 4 through 6) and by performing an aggregative, across-the-board mode of analysis which involved broad comparisons of relative prices, income, and total and disaggregated expenditures (chapters 7 and 8). A basic quantitative framework enabling us to treat these matters in a disciplined way appears in chapters 2 and 3. The central statistical construct of the study is the energy/GDP ratio, which is what remains after population is eliminated as the common element of energy consumption per capita and gross domestic product per capita.

A refrain has surfaced in energy discussions in recent years. It goes: "if the Swedes (or Germans or French or still others) can do it, why can't we?" and it is a blunt way of exposing this country's presumed disinclination to develop feasible and prudent energy-using habits. Although the conclusions of this study are tentative and their quantitative underpinnings too approximate to serve as the final word on this issue, both the organizing framework and the broad findings can strengthen public understanding. We have brought these findings together in this chapter in an integrated, interpretive setting. We discuss their implications and touch upon those topics which beg for further research.

Major Findings

Among the nine countries compared, the overall U.S. energy/GDP ratio is exceeded only by Canada. All other ratios are lower—ranging from 46 percent for France to 14 percent for Holland. Higher U.S. energy consumption relative to GDP is reflected to some extent throughout the range of energy uses. It is, however, much higher in some sectors of the economy than in others.

- Of four major sectors, there is least intercountry energy/GDP variability in the industrial sector: the unweighted average of energy consumption relative to output in Western Europe and Japan was only some 10 percent below that of the United States. If the energy sector is included, this gap grows to 20 percent.
- In the combined household and commercial sector, there was more variability, with a ratio averaging about 25 percent below that of the United States.
- Transformation sector losses—arising principally in the production of electricity—show a similar comparative spread.
- By far the greatest intercountry differences occur in the transport sector, where European and Japanese consumption of energy relative to GDP average 60 percent below the U.S. ratio. To put it another way, the transport sector of the U.S. economy uses more than twice as much energy as that of European nations and Japan for each dollar of national output.

A different way of pointing up these intercountry differences is by calculating the percentage contribution of the various sectors to differences in the aggregate energy/GDP ratio between the United States and each of the other eight countries. Thus, we find that on the basis of simple averaging, transport accounts for approximately half of the energy/GDP difference between the United States and the other countries (excluding Canada); the remainder of the difference is attributable to the household-commercial sector, about 17 percent; the industrial sector, about 23 percent (of which energy industries account for 16 percent and nonenergy industries, 7 percent); and transformation losses, 15 percent. A summary highlighting these differences appears in figure 9-1.

Another finding has to do with how expenditures for energy vary among countries. A revealing set of recently issued data on international

FIGURE 9-1   Energy/Output Ratios and Sectoral Composition, 1972

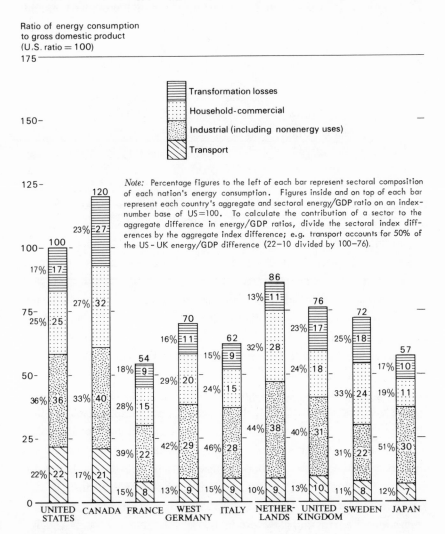

Ratio of energy consumption
to gross domestic product
(U.S. ratio = 100)

Transformation losses

Household-commercial

Industrial (including nonenergy uses)

Transport

Note: Percentage figures to the left of each bar represent sectoral composition of each nation's energy consumption. Figures inside and on top of each bar represent each country's aggregate and sectoral energy/GDP ratio on an index-number base of US=100. To calculate the contribution of a sector to the aggregate difference in energy/GDP ratios, divide the sectoral index differences by the aggregate index difference; e.g. transport accounts for 50% of the US – UK energy/GDP difference (22−10 divided by 100−76).

prices, income, and GDP is discussed in chapters 7 and 8 to shed light on this question. These data are expressed in dollars of comparable purchasing power and thus represent an improvement over international comparisons using market exchange rates as the basis of conversion. The use of market exchange rates leads to distortion in international price, income, and expenditure comparisons, and it is this distortion which the use of purchasing-power-parity rates of exchange is designed to minimize.

We found that in five countries included in this part of the analysis, the share of the consumer budget devoted to purchases of energy for household operations, and particularly personal automotive transport, is far below that in the United States. Spending for household fuel and power accounts for 3.2 percent of U.S. personal consumption expenditures, an average of 2.2 percent for four West European countries (France, West Germany, Italy, and the United Kingdom), and only 1.3 percent for Japan. The disparity for gasoline is much wider still. In addition, there are correspondingly wide intercountry differences in the respective proportions of the consumer budget accounted for by such energy-using durable goods as household appliances and cars. However, expenditures for purchased transportation, which is less energy intensive than passenger car use, are proportionately much higher—and by a wide margin—outside America. Underlying this difference is a clear-cut pattern of substantially higher prices abroad for fuels and power as well as for appliances and autos. In addition, higher overseas prices for automotive operation and lower prices for public transport provide added encouragement for using less energy-intensive transport modes.

One can make crude inferences about the effect upon aggregate energy/GDP ratios of these differential spending patterns. To make more than crude inferences requires looking at both the direct and indirect energy content of comparative patterns of expenditures.[1] Such analysis helps illuminate the effect on comparative energy/GDP ratios of distributional differences in spending patterns, on the one hand, and the energy associated with given expenditure categories, on the other. This is a distinction closely allied to a central concern of the study—that of separating the influence of energy intensity from compositional factors.

Proceeding along this line of inquiry, we start with the recognition that energy/GDP ratios in other countries may differ from the U.S. ratio because (1) the mix of final goods and services produced may be different from that of the United States, with a higher or lower proportion consisting of energy-intensive products; or (2) the mix may be roughly similar, but the energy needed to produce, transport, and distribute given products may be higher or lower than in the United States. Input–output analysis provides a means of capturing the effect that intercountry differences in the final-demand mix contribute to energy/GDP variability.

---

[1] "Direct" energy purchases are final-demand purchases not subject to any further processing or transactions, for example, household purchases of fuel for residential heating; "indirect" energy is what has been embodied in the manufacture or distribution of goods consumed, for example, the energy used to make aluminum beer cans.

With the U.S. 1970 energy/GDP ratio set at an index-number base of 100, the arithmetic average of the energy/GDP ratios for the five other countries comprising the input–output analysis comes to 57. About 19 percentage points of this 43-point difference (or, coincidentally, 43 percent of the total difference) are attributable to differences in composition of goods and services in other countries relative to the United States—largely this country's significantly higher consumer purchases of energy for household operations and gasoline for personal automobile transportation. Although other nations devote a proportionately greater share of their national product to capital investment expenditures (which tend to be relatively energy intensive), that characteristic does not even come close to offsetting the impact of U.S. household energy consumption.

The balance of 57 percent is due implicitly to two factors. Among these, the most important reflects different energy intensities among the countries for given categories of output, that is, variations in direct and indirect energy–output coefficients. In addition, the 57 percent share having been derived as a residual, the figure reflects an "interaction" effect attributable to both these factors.

These results were important in demonstrating, even before subsequent study provided strengthened evidence, that no unique set of circumstances can adequately explain intercountry energy/GDP variability. It would have taken a much more lopsided apportionment than the 43–57 split by which energy/GDP differences decomposed into final-demand mix, on the one hand, and intensities (along with the other factors), on the other, to warrant such a claim.

We must, however, add a qualifying word on interpreting the input–output analysis. What this analysis *does not* reflect is how efficiently fuels and power are deployed by the final user, which is an essential ingredient in interpreting these intercountry differences.

If, to use a simple example, an interindustry analysis were to reveal that the exclusive reason for country A's higher energy/GDP ratio was due to a larger volume of gasoline purchases than in Country B, the input–output finding would point to final-demand mix as the sole factor responsible. And that is conceptually the proper interpretation. However, suppose country A's larger volume of gasoline consumption were so offset by poor automotive fuel economy that it generated the identical number of passenger-miles as Country B. Clearly, if one were to go beyond the "mix versus energy-coefficient" analytical framework, one would find that the energy/GDP differences between A and B arose, in fact, from divergent efficiencies in the use of gasoline underlying the differential demand-mix patterns.

These thoughts are worth recording for the following reasons. We know—and our study bears out—that differences in economical utilization of energy by final users do figure in intercountry energy/GDP variability. Hence, if we were to allow for such end-use, energy-intensity differences *along with* the differences in energy intensity in the delivery of goods and services to final demand already accounted for, the 43–57 split would be altered. The mix factor would no doubt decline to some extent below 40 percent, and the coefficient factor would rise, although both for conceptual and measurement reasons, there is no precise way of attaching values to such an adjustment.

A somewhat different mode of analysis is used in chapters 4 through 6 to approach questions of mix and energy-intensity phenomena. Whereas the input–output approach focuses on the characteristics of final-demand expenditures, and treats these within a strict national accounts framework, these three chapters survey energy-using activities irrespective of point of final consumption, with less binding methodological constraints. Together with the review of national expenditures and the input–output analysis just described, the approach of chapters 4, 5, and 6 is designed to enrich the study by allowing us to treat our overall topic from varied perspectives and with differing degrees of detail, as dictated by the availability of data and the nature of given subject matter. Each broad segment of the economy is dealt with separately.

The sector comprising private households, commercial and public facilities, and agriculture accounts for one-fourth of U.S. energy consumption and averages a bit higher in the other countries. From an analytical standpoint, the burden of dealing with such a heterogeneous array of subsectors was mitigated by the fact that household uses dominate. Within the household segment of the sector, space heating (along with cooling in the United States) represents the leading energy use and received our primary attention. The significance of space conditioning within the household sector and—as it happens—its importance as a factor in overall energy/GDP variability is in itself worth noting for it implicitly relegates to a position of minor importance household energy-using gadgets and devices, which—particularly in their more frivolous manifestation—are sometimes cited as influencing unduly the high volume of American energy consumption.

After correction for climatic differences, the United States emerges as the largest consumer of space-conditioning energy relative to GDP. For numerous European countries, that ratio is 30 percent lower. It was estimated that about half this difference is attributable to the prevalence of larger, single-family homes in the United States. Most of the other half of the difference seems to arise from different heating habits, for example, the heating of unoccupied rooms or the insistence on higher temperatures, although these conclusions are derived in part from indirect evidence. For the residential uses in their entirety, the data suggest (1) a tendency for energy consumption to rise in line with, or slightly faster than, income across our range of countries and (2) a pattern of consumption that, within our limited field of observation, appears to be moderately responsive to fuel and power prices.

As noted earlier, comparative transport practices emerge as a major reason for the higher U.S. ratio. Not only are American passenger cars about 50 percent more energy intensive (in terms of passenger-miles per gallon) than European cars, but relative to given income levels, Americans also travel a lot more than Europeans. Indeed, this factor is somewhat more important than automotive energy intensity in explaining the far greater amount of energy devoted to transportation in the United States compared to Western Europe, relative to income. A third contributory element is the proportionately greater use in Europe of less energy-intensive public transport. We found these differences to exist not only because of the higher prices of acquiring and operating cars abroad—although that is clearly a significant element—but because of the urban density and public transport subsidies that exist abroad.

Freight transport also contributes to the higher U.S. energy/GDP ratio, but, interestingly, only because of the high volume of traffic (relative to GDP) that is generated in the United States, and not as the result of poorer fuel economy or an energy-intensive modal split. In fact, the U.S. freight modal mix is, more than Western Europe's, oriented to such energy-saving forms as rail, pipeline, and waterborne traffic. If one argues that size of country and long-distance shipment of bulk commodities (such as ores, grains, and coal) are inherent characteristics of U.S. economic structure and geography, one can make a case in which a relatively high U.S. energy/GDP ratio for freight is in no obvious way reflective of comparative inefficiency in the use of energy.

In our analysis of the industrial sector, we found that in the aggregate, the contribution of industry to higher U.S. energy/GDP ratios occurs in spite of the fact that the U.S. industrial sector is proportionately smaller than in the other countries. The United States exhibits more energy-intensive production processes. That is, if the individual value added were as large a proportion of U.S. national output as it is in most foreign countries, U.S. energy consumption would be even higher. What we observed for the industrial sector as a whole appears to be the case also in a diverse number of specific manufacturing segments. For example, the United States has a proportionately smaller iron and steel industry than a number of other countries, but a higher consumption of energy per ton of crude steel—a characteristic apparently arising from the prevalence of the high energy-using, open-hearth process and a relatively low prevalence of continuous rolling capacity. In some countries—notably Sweden—lower energy intensity in a number of product lines appears to result from the fact that technology is newer and economically optimized by virtue of considerably higher fuel prices.

Differences between the United States and other countries in industrial energy intensity need bear no clear-cut relation to the *overall efficiency* of carrying on a given industrial operation, such overall efficiency being a function of relative factor prices and quantity of other "inputs." To articulate the relationship between energy intensity and economic efficiency, it would be necessary to explore the cost of labor, capital, and nonenergy materials, in addition to the cost of energy resources. But the findings, as far as they go, constitute at least a strong presumption that U.S. industrial managers concerned with energy utilization might profitably explore the nature of foreign practices and results.

An analysis of the energy content of nonenergy commodities entering foreign trade showed that West Germany, Sweden, and Japan "export" about 10 percent of their energy consumption in the form of such embodied energy. Here we refer not to the trade in petroleum and other fuels themselves but to the energy that has gone into products such as steel. If we were to use an energy consumption measure adjusted for these "hidden" net outflows, the energy consumption of such countries would be still lower than recorded. For most of the countries (including the United States), embodied energy exports pretty much balance imports.

One final point regarding mix and energy intensity phenomena relates to comparative contributions to energy/GDP ratios arising from transformation losses in electricity generation (and, to a negligible extent, in other energy-conversion activities). The share of energy delivered in the form of electricity is not appreciably higher in the United States than elsewhere. In fact, it falls below the share of a number of countries. As a general proposition, therefore, and without evidence of markedly dissimilar electric generating efficiencies among countries, the higher U.S. energy/GDP ratio is, in part, due to disproportionately heavy reliance on electric power relative to national output, though not relative to other energy forms.

Interpreting the Findings

Can we extract from the range of sectoral findings a compact list of key factors operating on intercountry energy/GDP variability and—more important—indicate something about how much these key factors contribute to such variability? For this purpose, we limit our discussion to the United States in comparison with the six European countries, since some important elements in the energy/GDP comparison (freight transport, for example) could be dealt with only by using a six-nation European aggregate. We have, moreover, done some rounding and some arbitrary simplification of the quantitative complexity of the analysis.[2] The data in table 9-1 should therefore be viewed as an approximate rather than a precise formulation. Yet we have little doubt that these numbers convey the numerical essence of this study: namely, that there is no unique factor which, in quantitative terms, can be singled out as explaining the higher U.S. energy/output ratio. Rather, the 490 tons oil equivalent by which the United States exceeds the amount of energy associated with $1 million of GDP in Western Europe involves a variety of disparate factors.

Note, however, that four identifiable categories account for over 60 percent of the energy/GDP variability. In descending rank order they are: passenger transport, industry, residential space conditioning, and freight transport. Note also that in three of the sectoral cases shown, the contribution to total energy/GDP variability tabulated in the middle column represents the net item after a balance has been struck among factors pushing the U.S. energy/GDP ratio above the European ratio (those elements having positive numbers) and those depressing it below the European ratio (those with negative numbers). For example, in the

---

[2] An example is the rough apportionment of the "interaction" term arising in multiplicative relationships, such as in the case of coupling energy intensity with product mix. (Appendix E discusses this problem.)

TABLE 9-1   Contribution of Principal Factors to Higher Use of Energy Per Dollar of Output in the United States than in Western Europe, 1972

Sector or activity	Percent- age	Tons oil equivalent (per $ million GDP)
Total passenger transport	28	135
Volume of passenger mileage		58
Energy intensity		52
Modal split		24
Total freight transport	6	30
Volume of ton mileage		75
Energy intensity		−5 [a]
Modal split		−40 [a]
Total residential space conditioning	8	40 .
Size of units & prevalence of single-family dwellings		30
Heating practices		30
Degree day factor		−20 [a]
Total industry	20	100
Energy intensity		300
Structure		−200 [a]
Sum of above	62	302
All other (net)	38	188
Total energy/GDP variability [b]	100	490

Note: The West European figures underlying the tabulation represent a six-nation weighted average. Data may not add up to totals due to rounding. The breakdown in the third column is approximate.

[a] Negative numbers represent those elements depressing U.S. energy/GDP *below* the European ratio.

[b] The figure of 490 is the difference between the U.S. energy/GDP ratio (1,480 tons oil equivalent per $ million GDP) and the West European ratio (990).

case of space conditioning, the climatic adjustment means that, at the colder European temperatures, the U.S. ratio would have been still higher. Or with the European freight modal mix, the freight energy/ GDP ratio would have been much higher. Or, at the European industrial value-added share, U.S. industry's energy/GDP ratio would have been markedly higher.

What portion of the 490-oil-equivalent-ton difference between the United States and Western Europe can be assigned to structural factors, and what share to intensity features? In table 9-1 is a suggestive, though incomplete, answer. Somewhat arbitrarily, we can say that structural characteristics dominate higher U.S. energy/GDP ratios in the two transport components, while intensity features account for the higher U.S. industrial ratio. The split in household space heating is about half and half.

However, the unidentified part of the tabulation (188 tons oil equivalent, covering such activities as agriculture, commercial facilities, household uses other than space conditioning, and uses of energy in the form of a raw material rather than fuel [as in the case of asphalt]) is difficult to allocate. (Also, the presence of negative numbers in the last column of the table precludes, in principle, constructing a "clean" percentage distribution.)

But it seems to us, based on table 9-1, along with additional information, that the approximate intensity–structure (or "mix") division indicated by the final-demand analysis in chapter 7 would not be far off the mark for the sectoral analysis as well. The final-demand analysis shows mix accounting for 43 percent of the difference in aggregate energy/output ratios, and coefficients reflecting intensity responsible for 57 percent. We estimate that for the sectoral analysis, too, structural factors might explain about 40 percent of the overall difference, and intensity factors, the balance.

It therefore seems worth stressing that intensity factors alone—that is, higher amounts of energy per unit of activity or output in the United States compared to Europe—leave much of the energy/GDP variability to be accounted for by other characteristics of energy use. One reason this finding is of some importance is that it is perhaps easier (and less presumptuous) to visualize reduced energy-intensity differentials for particular outputs without negative effects on welfare than it is to conjecture about the implications of an energy-conserving change in the mix of final goods and services.

This identification of the major sources of intercountry energy/GDP variability is thus a useful and chastening first-order task. But such identification in essence is still only a statistical sorting out. It fails to treat the multilayered underlying factors that have given rise to the indicated differential patterns of energy consumption. Relative price differences are an obvious case in point, but even beneath relative price differences, there may be differences in, say, resource endowment, subsidies, or tax policies which price differences reflect. Or differential transportation patterns, which were seen to exert such enormous influence on energy/GDP differences, may—in addition to price factors and policy measures—reflect geographic, demographic, and industrial characteristics that constitute an additional set of determinants of variable energy/GDP relationships. At numerous points in the study, we have discussed various underlying phenomena having a presumptive effect on the observed energy/GDP differentials: for example, the way in which distances and low population density contribute to the high U.S. energy/ GDP ratio in transportation; or the role that the large size of dwelling units and other housing characteristics plays in shaping high U.S. residential energy use; or the impact of particular steel technologies on energy-intensity differences in manufacturing. These differences—to underscore a point made recurrently—may, but certainly need not, reflect efficiency differentials when the totality of productive factors used is considered along with energy. In the remainder of this chapter, we shall attempt to blend a few of these underlying features with the more visible quantitative characteristics already reviewed.

Economic Factors

At various points in the study we have drawn attention to the role of differing levels of prices in influencing consumption of fuels and power. This effect is particularly marked in both the transportation and the residential sectors. In the industry sector, where energy is only one among many inputs and typically a relatively small one in value terms, the influence of energy prices on consumption may be less pronounced, and in any case hard to isolate as a major determinant of energy-use patterns unless one considers the role of other inputs into productive processes. However, taking all these sectors together we find that the prices of fuels and power—ranging up to 100 percent higher in Europe and Japan than in the United States—are undoubtedly of considerable significance in explaining the 15–50 percent lower energy/GDP ratios overseas. Indeed, it would be surprising if consumption patterns were not to reflect differences in relative prices. Energy, in common with other things, is used in large volume when cheap, and more sparingly when expensive.

Because of a limited number of observations, it was not possible to examine the relationship between prices and energy consumption more systematically. But a recent study by William D. Nordhaus [3] covering similar countries gives price elasticities for each sector separately and

---

[3] International Institute for Applied Systems Analysis, "The Demand for Energy: An International Perspective," *Proceedings of the Workshop on Energy Demand* (Laxenburg, Austria, IIASA, May 1975) pp. 511–587. This study covers six of our countries—United States, France, West Germany, Italy, Netherlands, United Kingdom—but with a greater number of observations, as it includes pooled cross-sectional and time-series data for the period 1955–1972.

in the aggregate. Nordhaus reports a long-run price elasticity for aggregate energy consumption of -0.8. That is, a 10 percent increase in prices is associated with an 8 percent decline in energy consumption.[4] Although the fall in energy consumption is proportionately lower than the given price rise, it is nonetheless high enough—and, indeed, unexpectedly high compared with the findings of many other elasticity studies—to indicate that the relative prices of fuel and power, and of energy-intensive, compared with non–energy-intensive goods, are among the most important determinants of the energy intensiveness of the economy as a whole. In Nordhaus' words, ". . . it appears that relative prices play a crucial role in determining the energy intensiveness across space and time."[5] That judgment is probably correct even if "true elasticities" were half as high as Nordhaus' estimates.

Nordhaus also gives income elasticities, that is, the rise in energy consumption associated with increases in real income or output. Aggregate elasticity is about 0.8; that is, energy consumption increases proportionately less than the rise in income, at the same time being still strongly related to the rise in income.[6]

Together, these price and income elasticities explain much of the variation in levels of energy consumption in Nordhaus' study. While there is no reason to doubt consistency, in general terms, between the Nordhaus findings and our own analysis, the limited geographic and historical scope of the present study did not permit the kind of elasticity calculations that would establish such a corroborative link.

Other Factors

There are other considerations in addition to economic factors that obviously play some part in determining energy consumption, either

---

[4] Some estimated sectoral price elasticities are: domestic, -0.79; nonenergy industry, -0.5; and transportation, about -0.36. Prices refer to those paid in end-use consumption.

[5] International Institute for Applied Systems Analysis, p. 588. The powerful effect that an assumed price elasticity of demand even as low as -0.4 or -0.3 can have is shown in a recent study by the Oak Ridge Institute for Energy Analysis. (*Economic and Environmental Implications of a U.S. Nuclear Moratorium 1985–2010*, Oak Ridge, Tenn., August 1976.) There, such an aggregate elasticity, coupled with an assumed annual price rise of roughly 3 percent, yields an estimated future growth rate of U.S. energy consumption significantly below the rate prevailing during the historical climate of declining real prices. (This result obtains in spite of the assumption of a nearly 1.0 income elasticity of demand.)

[6] Income elasticies for selected sectors are: domestic, 1.09; transportation, 1.34; nonenergy industry, 0.76. *Individual* country experience may, of course, depart from these generalized findings. For example, analysis of American families' energy use by income classes suggests lower income elasticity.

by working through economic variables or independently. We shall touch upon four: geography, energy resource endowment, import dependence, and fuel mixes.

GEOGRAPHIC CHARACTER. The geographic character of a country can affect energy consumption in several ways. Climatic differences, for example, have been shown to influence the energy consumption of both Canada and Sweden substantially, although, it will be recalled, climate plays no part in explaining the higher U.S. consumption compared with other countries. Similarly, the size of a country can affect energy consumption because of varying needs for mobility in transportation and, less important, because of higher gas and electricity transmission losses in larger countries.

ENERGY RESERVE ENDOWMENT. In addition, the amount of energy resources a country has can affect energy consumption. The possession of energy resources tends to increase energy consumption in two ways. First, the energy-extractive industries, particularly natural gas, require large inputs of energy in production, processing, and delivery of the product. In three countries with large energy resources—United States, Canada, and the Netherlands—there is a much higher consumption of energy in the energy-producing sector relative to income, compared with the other countries. Second, the existence of energy resources may encourage the development of energy-intensive practices and activities based on the availability of cheap energy. Thus, both Sweden and Canada, with proportionately the greatest hydro capacity, have important electro-process industries and, in addition, show the largest contribution of electricity to industrial energy use. In the case of coal, with its stimulus toward development of iron and steel industry, such energy-intensive industrial characteristics may endure even when the original resource conditions have changed.

With regard to the link between abundant energy resources and lower energy prices, however, it is important to note that intercountry experience varies considerably depending on the energy source or form in question. Natural gas appears to be a comparatively cheap form of fuel in all countries where it is produced. But, depending on the country, coal may not be. Among the nine countries, only the United States and Canada were significant crude oil producers in 1972, but the difference between the generally lower U.S. prices for petroleum products, compared to those in Europe and Japan, arises mainly from differences in taxes. These have persistently been much higher in Western Europe. The presence of large energy resources in a country thus does not automatically ensure the lowest prices, although, in general, there is a tendency for those countries with abundant energy resources to consume more energy relative to income than the others.

IMPORT DEPENDENCE. The other side of the coin is a lack of domestic resources, which involves dependence on imported supplies. The degree of energy import dependence (measured by the percentage of net imports in consumption) varies considerably among the nine countries. Only one, Canada, was a net exporter in 1972 (to the extent of 12 percent of consumption). The United States experienced a relatively low degree of energy import dependence—14 percent of consumption in 1972. The other countries displayed much greater import dependence, varying from 40 to 50 percent in the case of West Germany and the United Kingdom, to between 70 and 90 percent for France, Italy, Sweden, and Japan.

As figure 9-2 shows, the association between energy import dependence and energy consumption relative to GDP is remarkably close. This connection may come about in a number of ways. The absence of energy resources means that there are fewer energy industries themselves, and these, as in the case of petroleum production, tend to be rather energy intensive. But there is no reason to believe that absence of energy resources has necessarily discouraged development of other energy-intensive manufacturing industries, or that depletion of domestic resources has precluded the continuation of energy-intensive industries. Japan, for example, consumes as much energy in industry (relative to national output) as the other, less dependent, countries. As we demonstrate in chapter 6, much of the output of those energy-intensive industries is destined for export rather than domestic consumption. Consequently, costs of imported energy are recovered in exports. Germany has a strong historic base for such a structure in the form of its coal industry.

A related link between import dependence and energy consumption lies in the energy price levels of the more import-dependent countries, which tend to be higher than those of the less import-dependent countries. Part of this difference is because of the higher taxes levied on energy and, particularly, on petroleum products. In many European countries, these have tended to account for 50 percent of gasoline and about 20 percent of heating oil and heavy fuel oil prices. By contrast, in

FIGURE 9-2   Energy Import Dependence Compared with Energy/ Output Ratios, 1972

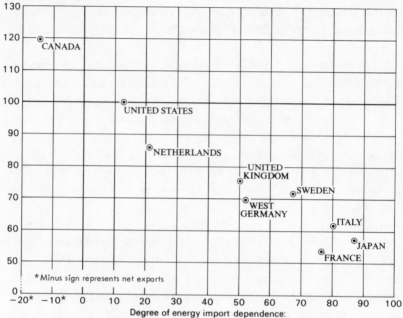

Ratio of energy consumption to gross domestic product (U.S. ratio = 100)

Degree of energy import dependence: net energy imports as a percentage of total energy consumption

the United States, taxes have accounted for only about 20 percent of gasoline prices, and there have been no excise taxes levied on heating fuel or heavy fuel oil. Undoubtedly, a variety of motives—including preclusive ones like governmental revenue generation and protectionism—lay behind the tax policies of the European countries; but in the highly import-dependent countries, where fuel imports are a substantial proportion of the import bill, such measures can at least partly be attributed to concern over the balance of payments and efforts to limit dependence on imported supplies.

FUEL MIXES. These variations in energy resource endowment and in import dependence result in very different fuel mixes among the countries. Petroleum products are the single most important source of energy for all countries, providing between one-half (for the United States, Canada, West Germany, Netherlands, and the United Kingdom) and 65 to 75 percent (for France, Italy, Sweden, and Japan) of energy consumption. For the other fuels: coal provides 30 percent of total West

German and U.K. consumption; gas, 30 to 40 percent of U.S. and Netherlands consumption; while "primary" electricity (hydro and some nuclear) is a major source (24 to 30 percent) only for Canada and Sweden. At the final consumption stage, although petroleum products remain the single largest energy form, providing over one-half of the total, electricity becomes much more important, providing for most countries 10 to 15 percent of the total if heat losses are excluded, and about 25 to 30 percent, including heat losses.

The nature of the different fuel mixes may impinge on overall energy intensity because of the varying thermal efficiencies of the different fuels. Liquid fuels and gas have a higher thermal efficiency than coal—particularly in household use where coal burned in fireplaces has only one-third the efficiency associated with the other fuel heating systems. In the household and industrial sectors, however, both the United States and Canada have a relatively high thermally efficient fuel mix compared with most other countries. Thus, the higher North American consumption level (relative to output) compared with other countries is not explained by differences in fuel mix. Among the other countries, only the United Kingdom seems significantly out of line, having relatively low thermal efficiency stemming from the use of coal in open grates for domestic heating. Fuel mix, therefore, explains part of the United Kingdom's relatively high energy consumption relative to GDP.

An Energy-Intensive Society

In the light of the topics covered by this study, it is possible to sketch out the principal ingredients characterizing an energy-intensive economy. Compared to the ratio elsewhere, a country's energy/GDP ratio will be higher

—the lower its relative fuel and power prices and costs

—the larger its passenger-mile volume relative to GDP

—the poorer the fuel economy of its passenger car fleet

—the lower the proportionate role of public transport

—the larger its freight-ton-mile volume relative to GDP

—the lower the proportionate role of rail, pipelines, and waterways in freight transport

—the larger the size of housing units

—the larger the share of single dwelling units

—the colder the climate

—the greater the proportionate importance of extractive and manufacturing industry

—the more energy intensive the industrial sector

—the greater the proportionate role of electricity

—the greater the degree of self-sufficiency in energy supplies.

In table 9-2, we have ranked our nine countries with respect to certain of these criteria. The table is based on data and materials developed in various chapters of this study. Needless to say, some criteria—being essentially tautological ("the more energy intensive the industrial sector")—are listed only to leave with the reader a feeling for the significant factors emerging from our analysis. Further, there is obviously overlap among the items; for example, lower prices contribute to indifference about poor gasoline mileage.

A ranking of countries according to these characteristics shows that the United States, along with Canada, heads the list in most, but not all, instances. (We had no country-by-country freight data, but found the United States to be characterized by lower energy intensity and an energy-saving modal mix.) The pervasively high North American rankings, correlating with high energy/GDP ratios, is accompanied in these two countries by the lowest energy prices.

But it is worth bearing in mind, finally, that the comparative energy patterns disclosed and analyzed in this report derive essentially from a cross-sectional picture snapped at a point in time. While many of the characteristics shaping energy/output differences in the early 1970s probably reflect enduring features of the different economies, it is also true that a study of historical changes—both prior to and following 1972—makes us see things in a slightly different way, and, as a result, perhaps qualify them. For example, figure 1-4 in the first chapter shows that during the past decade and a half energy/GDP ratios have moved distinctly higher in the Netherlands and Sweden, in contrast to a much more sidewise movement in the United States. The widespread introduction of natural gas for central heating and a rapid rise of automotive travel seem to have been factors at work in the former countries respectively. Similarly, events since the 1973–1974 oil disruption—including a marked escalation in energy prices and greater or lesser adoption of public conservation policies—are apt to induce energy consumption responses whose effect may well be to reshape the differential patterns

TABLE 9-2   A Schematic Ranking of Factors Affecting Comparative Energy Consumption/GDP Ratios, by Country, 1972

(for a given factor, the lower the number, the greater the effect on raising the energy/GDP ratio)

Factors	United States	Can- ada	France	West Ger- many	Italy	Nether- lands	United King- dom	Sweden	Japan
Energy prices (lowest prices = 1)	1	2	4	5	9	6	8	3	7
Passenger-miles per unit GDP	1	5	9	3	2	6	4	6	8
Percentage of passenger-miles accounted for by cars	1	2	8	4	7	5	6	3	9
Energy consumption per car-passenger-mile	2	1	4	8	5	7	9	6	3
Cold climate [a]	7	2	5	4	9	3	5	1	8
Size of house & percentage single family	1	1	6	6	8	5	4	3	8
Extractive industry GDP as percent of total GDP	2	1	n.a.	4	6	n.a.	3	5	7
Industrial GDP as percent of total GDP	7	8	2	1	6	4	5	9	3
Ratio of industrial energy consumption to industrial GDP	2	1	9	8	6	5	4	3	7
Degree of energy self-sufficiency	2	1	7	5	8	3	4	6	9
For reference:									
Energy/GDP ratio	2	1	9	6	7	3	4	5	8
Energy per capita	2	1	7	5	9	4	6	3	8
GDP per capita	1	3	4	5	9	6	8	2	7

Note: When number is missing in rankings, it is due to a tie; n.a. = not available.

Sources: Based on data appearing in chapters 3 through 8 of this study, supplemented in the case of line 7 by information from United Nations, Statistical Yearbook, 1974, New York, 1975.

[a] Measured by degree days.

revealed in our study. Witness, for example, the modest momentum toward enhanced automotive gas economy now under way in the United States.[7]

⌒◦

Upon reflection, some of the remarks made in this chapter may convey a narrow appreciation of the latitude for transferring energy-conserving practices from one high-income country to another—*given* differences in prices, structure, geography, resource endowment, population characteristics, tastes, and other factors. After all, if economic logic dictates that U.S. energy prices—whether low because of regulatory controls, market conditions, or an absence of taxes—favor a more energy-intensive manufacturing process or less energy thrift in household operations than would be the case with prices at relatively high foreign levels, it is easy to wind up with the somewhat facile conclusion that rationality operates and all is for the best.

This is decidedly not our message. For one thing, some of these characteristics now show clear signs of changing—either propelled by external circumstances (for example, higher market prices) or, prospectively, by policy measures (for example, taxes or subsidies) to restrain demand. It is precisely this new environment that makes it useful to ponder the potential benefits to the United States of energy management—be it in households, industry, or transport—under motivations that have traditionally prevailed elsewhere but may now begin to be applicable here. For example, shifts toward substantially enhanced fuel economy in passenger cars and improved heating practices could begin giving the United States some of the energy-using attributes of Western Europe. These two categories alone could probably reduce total U.S. energy consumption by about 7 percent and shave the U.S./European differential (shown in the bottom of table 9-1) by 15 to 20 percent. A modest move toward more public transport could strengthen such a shift. Although we hesitate to pronounce such steps as inconsequential for life-styles (what does or does not cramp peoples' life-styles is a legitimate matter of definition and debate), these are steps that, in our judgment, are eminently feasible. They are not about to threaten the social fabric.

Prospective increases in real energy prices are likely to be a persuasive inducement for more energy-conserving practices. Noting that under

[7] In a current Resources for the Future study, we take up the nature of these historical developments—both preceding and following the 1973–1974 events—as they bear on comparative intercountry patterns of energy relative to national output.

prevailing energy price levels, potential U.S. energy savings are at least in part precluded by an absence of economic incentives, a study by the International Energy Agency (IEA) recently observed: "One particularly disturbing fact . . . is that some IEA Members continue to price energy below world market levels. This can result in diminished future investments in energy efficiency and less attention by energy consumers to the wise use of energy." [8] To U.S. policy makers or advisers intrigued by more sparing foreign energy use, the IEA quotation, consistent with a major thrust of this study, serves as a useful reminder: energy consumption patterns are not shaped in a total policy vacuum; to the extent that they derive from particular pricing, fiscal, and other economic-policy measures, there exists a burden of considering the acceptability of such policy instruments in the United States in order to encourage similar patterns here.

Of course, there are cost-effective avenues toward conservation that can be deployed to complement and reinforce purely price-induced restraint. Mandated gasoline economy standards for cars, enforced building insulation practices, and guidelines for appliance efficiency are particular areas for potential energy savings. For example, when combined with mortgage allowances for energy-conservation investments, Swedish weatherization standards for buildings have contributed to economizing in the use of heat. In a number of foreign countries, thermal generation of electric power combines steam heat and electricity for community "district heating" schemes or for industrial "co-generation" of process steam and electric power. Although the configuration of U.S. locational patterns may not be suitable for extensive district heating—for example, because of heat losses in trying to serve dispersed consumption centers—potential energy savings from this practice are sufficiently large that its possible use in the United States deserves a close look. A number of institutional, regulatory, environmental, and technical questions need to be dealt with before widespread recourse to industrial co-generation occurs in the United States; such questions are beginning to receive prominent attention.

But, in the interest of balance and perspective, we should keep in mind that there are practical limits to the opportunities for instituting foreign-inspired energy husbandry in the United States. The United

[8] Organisation for Economic Co-operation and Development, *Energy Conservation in the International Energy Agency: 1976 Review* (Paris, OECD, 1976) p. 8.

States is not about to acquire the population density of Japan, the geographic compactness of Germany, nor—very likely—the rail network of France, so that in the best of circumstances, some intercountry energy/GDP variability is bound to endure because of historical or deeply rooted reasons rather than because of false energy price signals or misguided energy management. Social customs and conscious individual preferences having different energy-use implications will persist as well. Moreover, energy-saving measures are at least equally likely to be pushed in other countries; so the United States will find itself aiming at a moving target.

Even so, it is unquestionably true that the longer the passage of time, the greater the opportunities for substituting energy-saving processes, equipment, and even life-styles, for energy-intensive practices and capital. In that longer run sense, what now seems "structurally" frozen may turn out to be quite flexible.

No tidy final reckoning is possible on these questions. Our study points to complex and diverse reasons for intercountry differences in energy consumption. Variations in energy/output ratios should not in themselves be viewed as indicators either of economic efficiency or even of energy efficiency. Economic efficiency depends on how energy is used in combination with other resources—particularly capital and labor—and the relative costs of all of these. National energy/output ratios also depend critically on the composition of a country's output, and not merely on the energy intensities associated with these component products and activities.

Some of these compositional differences, such as automotive patterns, appear to be decisively influenced by relative user costs. Other notable differences, such as suburbanization, housing, and mobility characteristics—each with marked influence on energy use—might arguably owe some of their historic momentum to cheap energy in the United States and costly levels elsewhere, but are clearly related to many other impelling forces as well. Whatever their origins, some of these features have become such firmly established elements in American society that it would be unrealistic to expect dramatic change or reversal—even in a new energy-price era. Viewed in this way, in numerous of its aspects, energy consumption is essentially a by-product or, at best, only one element within the wider framework of societal arrangements and choices.

This has been a study whose primary objective has been to identify and sort out some of the principal factors that account for differential energy patterns. The more deep-seated issues touched on in the last few pages—and especially the clear-cut policy lessons for the United States— deserve still deeper probing and discussion. It is nevertheless our firm judgment that, in the comparative intercountry setting of this report, the assumed presence of an energy-conservation ethic and abhorrence of waste "over there," in contrast to the disregard for such things in the United States, is simply not supported by the facts and is simplistic in its view of the world. The need to proceed with energy conservation strategies appropriate to the United States is too urgent a task to allow ourselves to be too much transfixed by a foreign yardstick that, although intermittently revealing, can also be illusory.

Appendixes

Derivation of Gross Domestic Product Estimates

TO MAKE INTERCOUNTRY COMPARISONS of total output it is necessary to convert the gross domestic product (GDP) of each country as it
is expressed in its own currency into comparable units, usually U.S.
dollars. There are two broad alternatives—the use of market exchange
rates and the use of estimated exchange rates reflecting purchasing power
parities.

The purchasing-power-parity (ppp) approach is superior on theoreti-
cal grounds, but has been little used in the past owing to lack of the
necessary detailed country data. We have been fortunate in this project
in benefiting from a book published in 1975.[1] This book will hereafter
be referred to as the ICP study or the *ICP Report*. This report includes
four purchasing power parities, based on four different sets of weights for
five of our countries—France, Germany, Italy, United Kingdom, and
Japan, all applying to the year 1970. To adapt the ICP material to our
needs we had therefore to:

1. choose which of the purchasing power parities were the preferred
   ones for our purpose

2. update them to our reference year, 1972

3. make estimates for Canada, Sweden, and the Netherlands, which
   were not covered in the original ICP study.

Adaptation of ICP Material

The ICP study gives four parities applying to GDP for each of the five
U.S.–other country comparisons. These are three binary parities based
on U.S. quantity weight, other-country quantity weight, and "ideal"
weight (which is the geometric mean of the U.S. and other country weight

---

[1] Irving B. Kravis, Zoltan Kenessey, Alan Heston, and Robert Summers, *A System of International Comparisons of Gross Product and Purchasing Power* (Baltimore, published for the World Bank by The Johns Hopkins University Press, 1975).

TABLE A-1   Market Exchange Rates and Purchasing Power Parities, 1970

Parities	France Fr/$	W. Germany DM/$	Italy L/$	U.K. £/$	Japan Y/$
Market exchange rate	5.554	3.66	625	0.417	360
Binary parities					
U.S.-weighted	4.95	3.49	518	0.324	275
Country-weighted	4.10	2.92	414	0.274	220
Ideal	4.51	3.19	463	0.298	246
Multilateral parities	4.48	3.14	483	0.308	244

Source: Irving B. Kravis, Zoltan Kenessey, Alan Heston, and Robert Summers, *A System of International Comparisons of Gross Product and Purchasing Power* (Baltimore, published for the World Bank by The Johns Hopkins University Press, 1975), tables, series B, pp. 171, 172, 175, 176, and table 14.3, p. 233.

indexes); and a multilateral parity in which each country's quantity weights are used with international prices based on prices from a range of countries, including developing countries. The different parities, together with the market exchange rate of that year, are in table A-1. In all cases, all four parities were below the market rate for the U.S. dollar, implying that the levels of other countries' GDP converted by market rates would be undervalued, in some cases, substantially. Between the various parities, the U.S.-weighted parity tends to be closest to the market rate, and the country weight furthest away.

With regard to the choice between these parities, the multilateral had at first sight the greatest advantage, as it alone would have permitted comparisons between third countries. The other rates, being binary, strictly speaking only permitted comparison between any two countries, of which one was the United States. Thus using binary parities, it could be said that in index terms of U.S. = 100, Germany has a per capita income of 70 and Italy, 45. But these comparisons would not, strictly speaking, permit any conclusions to be made about the difference between the levels of Italian and German incomes.

But we chose the binary rather than the multilateral because: (1) being based on common prices, the multilateral would have made it more difficult for us to get at price differences between countries—which we felt were certain to play a significant role in differences in energy consumption; (2) the prices chosen in the multilateral are based on a wide range of countries, including, for example, Kenya and India whose economies differ markedly from the economies of the countries covered in our study; and (3) a bridging exercise done in the ICP study gave reason to believe that for our countries and our purposes, third countries linked in a binary fashion could in practice be compared without being too far off the mark.

Among the binary parities, the ideal was chosen because it has the advantage of avoiding the extremes of both U.S. and the other country's weights.

Having decided on which parity to use, the next step was to update the binary ideal to 1972. Two procedures were considered. The first was to use the binary ideal to deflate the 1970 values of GDP (ICP adaptations of UN data) in own-country currency into U.S. dollars, and then increase the GDPs by the percentage increase in real product which had taken place in these countries from 1970 to 1972.

An alternative method was to increase the binary ideal parity by the country's price rise relative to the U.S., and deflate 1972 OECD data in own-country currencies by the adjusted binary ideal parity. It was decided to use the updated parity approach on the grounds that: (1) it facilitated the estimation of parities for countries not covered in the ICP study and for which 1970 GDP data in ICP terms were not available; and (2) it would provide us with parities for purposes not directly associated with GDP, such as the deflation of price data. In practice, the results using the two alternatives were very similar and would have a negligible effect on the relative standing of the nine countries' energy/GDP ratios.

Finally, estimates had to be made for those countries not covered in the ICP study—Canada, Sweden, and the Netherlands. For Canada the market exchange rate (the 1970 rate adjusted upward by the price rise relative to the United States) was felt to be an adequate approximation of purchasing power parity on the grounds that the closer the levels of GDP between two countries, the less the distortion introduced in valuing output by market exchange rates.[2] More precisely, an article by Daly and Walters shows (for 1960 at least) considerable similarity between the valuation of Canadian and U.S. output whether using purchasing power parities or exchange rates.[3] (Note that unlike the other countries there is virtually no difference between Canadian output valued at U.S. price weights and Canadian output valued at Canadian price weights.)

For Sweden too, the similarity in income levels gave some presumption that market exchange rates would be a reasonable approximation of pur-

[2] Paul A. David, "How Misleading Are Official Exchange Rate Conversions?" *Economic Journal* vol. 82, no. 1 (September 1972) pp. 979–990. The article includes references to literature on this topic.

[3] D. J. Daly and D. Walters of the Economic Council of Canada, "Factors in Canada–United States Real Income Differences," Paper presented at the International Association for Research in Income and Wealth, Maynoth Conference, Ireland, August 1967.

chasing power parities. At the same time, there was clearly more difference between relative prices in the United States and Sweden than there was between United States and Canada, so that rather than take the market exchange rate as it stood, a binary ideal parity for 1970 was interpolated on the basis of observed relationships among the different rates for each of the other countries (see table A-1). The 1970 estimate was then updated to 1972.

For the Netherlands, where income differences are greater, purchasing power parities were derived by interpolation as follows. A further extension of the ICP study gave the exchange rate deviation index for consumption (presumably multilateral) for the Netherlands, which yields a consumption ppp ratio of 2.78 guilder to the dollar.[4] From the ICP study it appears that for other, like European countries the consumption ratio is similar but somewhat higher than the GDP ratio. The ratio of 2.78 was therefore reduced to 2.7 in line with other countries. As the multilateral rate for other countries was found to be similar to the binary ideal, 2.7 was also taken as the binary ideal and updated to 1972 (3.0).

Results of Estimates

The results of these adjustments and estimates are in table A-2. They show parities for 1970, adjusted parities for 1972, and the exchange rates of these years. Note the sharp movement in exchange rates between these years associated with floating after the 1971 Smithsonian Agreement. In almost all cases this was a movement toward the appreciation of foreign currencies relative to the dollar, and, as such, in the direction of purchasing power parities. This effect is illustrated by the deviation of the purchasing power rate from the exchange rate (expressed as indexes in the final column of table A-2). By 1972, the market rates for several countries had come very close to the binary ideal parity, although significant deviations still prevailed for Italy, United Kingdom, and Japan.

The sensitivity of GDP levels to the choice of deflator is shown in table A-3. For France and Germany the difference is relatively small, but for the other three countries, more substantial. The choice of the binary, in practice very similar in level to the multilateral, and falling, by

---

[4] I. B. Kravis, A. Heston, R. Summers, and A. Civitello, "Three Phases of the International Comparison Project," Paper presented at the 14th General Conference of the International Association for Research in Income and Wealth, A-ulanko, Finland, 1975.

TABLE A-2   Exchange Rates and Purchasing Power Parities, 1970 and 1972

Countries	Binary ideal parity	Exchange rate	Index of exchange rate deviation (U.S. = 100)
	(foreign currency per $)		
	1970		
Canada (C$)	[1.01]	1.01	[100]
France (Fr)	4.51	5.52	122
W. Germany (DM)	3.19	3.65	114
Italy (L)	463	623	135
Netherlands (g)	[2.70]	3.60	[133]
U.K. (£)	0.298	0.417	140
Sweden (Kr)	[4.63]	5.17	[111]
Japan (Y)	246	358	146
	1972		
Canada (C$)	[1.007]	0.995	[99]
France (Fr)	4.65	5.13	110
W. Germany (DM)	3.37	3.20	95
Italy (L)	483	582	120
Netherlands (g)	[3.0]	3.2	108
U.K. (£)	0.322	0.426	132
Sweden (Kr)	4.90	4.74	97
Japan (Y)	250	302	121

Note: Figures in brackets are RFF estimates.

Source: ICP Report data updated by RFF. 1970 figures (in brackets) are RFF estimates.

ᵃ Index of exchange rate deviation is exchange rate divided by the binary ideal parity.

definition, within the limits of the U.S.- and country-weighted totals, increased the level of income relative to the United States for all countries except West Germany. In table A-3, the levels of GDP yielded by the application of the different parities and the market exchange rate to 1972 data are only approximate. The deflation of country data by exchange rates reflecting purchasing power parities is intended to improve upon the recognized deficiencies of market exchange rates as a means of deflation. Nonetheless, the results must still be treated with caution.

In terms of the energy/GDP ratios for our countries, therefore, the effect of using market rates compared with *any* purchasing power parity is to compress differences in energy/GDP ratios. That is to say that any parity adjustment indicates that the differences between the energy/GDP ratio of the United States and other countries (Canada apart) are

TABLE A-3   Sensitivity of 1972 GDPs to Various Purchasing Power Parities

(billion dollars GDP)

Countries	Market exchange rates	Binary parities			Multilateral parity
		U.S.-weighted	Country-weighted	Ideal	
U.S.	1,178	1,178	1,178	1,178	1,178
Canada	106	—	—	[103]	—
France	196	196	237	215	217
W. Germany	258	225	268	246	250
Italy	118	127	159	142	136
Netherlands	46	—	—	[49]	—
U.K.	154	175	207	190	184
Sweden	41	—	—	[41]	—
Japan	294	325	405	362	366

Note: Blanks indicate no estimate was made. Figures in brackets are GDPs based on exchange rates estimated by RFF.

Source: ICP Report data updated by RFF.

greater than that given by an energy/GDP ratio based on market exchange rates.

Binary Comparisons and Bridging

As noted previously, the use of data based on the binary ideal parity is limited to direct comparisons between two countries, one of which is the United States. Clearly it would be very useful if in addition we could make third country comparisons even though the emphasis in the report is not primarily on comparing third countries.

This question was addressed in the *ICP Report*. Comparisons of third country GDPs with each other through the United States (bridging) were made on the basis of detailed category information, as contrasted with the third country comparisons at GDP level in our binary comparisons. Bridging at the detailed category level has the advantage of reducing, though not eliminating, the effect of the bridge country (the United States), because it has no effect on the weights used in aggregation. The results of these two bridging exercises are given in table A-4. Base countries appear in the left-hand column. To take an example: with France's GDP per capita at 100, then Italy's GDP is 64.2 of France's if bridging is done at the GDP level, and 63.0 if bridging is done at the detailed cate-

TABLE A-4 Bridging of GDPs at Aggregate and Detailed Category Levels, 1970

(U.S. = 100)

	Numerator country			
Base country	Italy	Japan	U.K.	W. Germany
Italy	—	—	—	—
Japan				
GDP level	78.4	—	—	—
Detailed category level	84.2	—	—	—
U.K.				
GDP level	76.5	97.6	—	—
Detailed category level	77.3	95.3	—	—
W. Germany				
GDP level	64.9	82.9	84.9	—
Detailed category level	64.9	79.8	83.7	
France				
GDP level	64.2	81.9	83.9	98.8
Detailed category level	63.0	77.2	80.7	98.1

Note: Blanks = not applicable.
Source: ICP Report, adapted from table 4-1, panels A and B, p. 51.

gory level. The closer the two ratios, the greater the justification in using binary data linked at the GDP level for third country comparisons.

For the countries in table A-4, the two ratios generally coincide. The greatest difference occurs when Japan is compared with other countries, indicating that greater caution should be used in comparisons that include that country. Otherwise the table offers adequate justification for making third country comparisons even though the basic data are in binary terms.

Basic Statistical Tables

THE TABLES in this appendix were derived from the following sources:

- for population data—Organisation for Economic Co-operation and Development, *Labor Force Statistics, 1962–72* (Paris, OECD, 1973)

- for gross domestic product (GDP)—national accounts data appearing in Organisation for Economic Co-operation and Development, *National Accounts of OECD Countries, 1961–73* (Paris, OECD, 1974)

- for energy consumption—data appearing in Organisation for Economic Co-operation and Development, *Statistics of Energy, 1958–72* (Paris, OECD, 1973) were converted into tons oil equivalent (toe). An account of the conversion factors used and the method of arranging converted data into country balance sheets are available from the Center for Energy Policy Research of Resources for the Future.

TABLE B-1 Basic series: Total Energy Consumption, Gross Domestic Product, and Population, 1972

Item	U.S.	Canada	France	W. Germany	Italy	Netherlands	U.K.	Sweden	Japan
Total energy consumption (million tons oil equiv.)	1,744.65	183.09	171.23	253.85	130.01	62.36	213.10	43.19	307.82
Gross domestic product ($ billion)	1,178.49	103.31	215.47	246.10	142.13	49.03	190.03	40.65	362.71
Population (millions)	208.84	21.85	51.70	61.67	54.41	13.33	55.88	8.13	105.97
Energy consumption Per $ million GDP	1,480	1,712	795	1,031	915	1,272	1,121	1,062	849
Index (U.S. = 100)	100	119.7	53.7	69.7	61.8	85.9	75.8	71.8	57.3
Energy consumption Per capita (tons oil equivalent)	8.354	8.379	3.312	4.116	2.389	4.678	3.814	5.312	2.905
Index (U.S. = 100)	100	100.3	39.6	49.3	28.6	56.0	45.6	63.6	34.8
GDP per capita (dollars)	5,643	4,728	4,168	3,991	2,612	3,678	3,401	5,000	3,423
Index (U.S. = 100)	100	83.8	73.9	70.7	46.3	65.2	60.3	88.6	60.7

219

TABLE B-2 *Energy Balances, 1972*
(million tons oil equivalent)

Item	U.S.	Canada	France	W. Germany	Italy	Netherlands	U.K.	Sweden	Japan
Production	1,524.95	216.82	45.82	125.77	25.27	48.92	103.50	13.69	46.73
Imports	275.88	61.08	149.47	159.76	140.07	80.22	134.45	32.57	282.42
Exports	39.88	84.00	12.86	21.87	27.30	54.27	20.53	1.90	2.40
Stock changes[a]	−3.18	−7.59	−5.60	−4.05	+0.58	+0.27	+2.08	+0.49	−7.12
Total	1,757.78	186.31	176.83	259.61	138.61	75.14	219.50	44.85	319.63
Bunkers	13.13	3.22	5.60	5.76	8.60	12.78	6.40	1.66	11.81
Total energy con- sumption	1,744.65	183.09	171.23	253.85	130.01	62.36	213.10	43.19	307.82
Net imports as percent of total energy consumption	12.7	−14.3	76.5	52.1	80.1	21.1	50.5	67.2	87.1

[a] Minus sign = stock accumulation; plus sign = stock withdrawal.

TABLE B-3 *Energy Consumption by Sector, 1972*
(absolute figures in million tons oil equivalent)

Sectors	U.S.	Canada	France	W. Germany	Italy	Netherlands	U.K.	Sweden	Japan
Total consumption	1,744.65	183.09	171.23	253.85	130.01	62.36	213.10	43.19	307.82
Transformation losses	294.53	41.47	30.22	41.83	18.96	8.03	48.24	10.86	53.46
Percent of total	16.9	22.7	17.6	16.5	14.6	12.9	22.6	25.1	17.4
Energy industry	158.55	13.24	12.27	18.30	6.79	4.89	15.32	1.35	17.57
Percent of total	9.1	7.2	7.2	7.2	5.2	7.8	7.2	3.1	5.7
Transport	385.11	31.55	25.30	32.59	19.36	6.57	27.49	4.91	38.09
Percent of total	22.1	17.2	14.8	12.8	14.9	10.5	12.9	11.4	12.4
Industrial	364.40	40.04	47.17	73.62	40.05	12.47	60.52	11.17	119.71
Percent of total	20.9	21.9	27.5	29.0	30.8	20.0	28.4	25.9	38.9
Household-commercial	440.85	49.56	48.13	73.92	31.25	19.96	51.53	14.15	59.51
Percent of total	25.3	27.1	28.1	29.1	24.0	32.0	24.2	32.8	19.3
Nonenergy use	101.20	7.24	8.13	13.58	13.60	10.44	10.01	0.75	19.50
Percent of total	5.8	4.0	4.7	5.3	10.5	16.7	4.7	1.7	6.3

TABLE B-4 Energy Consumption by Sector, Including Heat Losses, 1972
(absolute figures in million tons oil equivalent)

Sectors	U.S.	Canada	France	W. Germany	Italy	Netherlands	U.K.	Sweden	Japan
Total consumption	1,744.65	183.09	171.23	253.85	130.01	62.36	213.10	43.19	307.82
Energy industry	205.00	17.77	16.32	25.65	10.08	5.95	22.86	2.76	25.72
Percent of total	11.8	9.7	9.5	10.1	7.8	9.5	10.7	6.4	8.4
Transport	385.82	31.64	26.27	33.96	19.95	6.72	27.92	5.23	40.08
Percent of total	22.1	17.3	15.3	13.4	15.3	10.80	13.1	12.1	13.0
Industrial	486.25	57.11	59.80	93.58	51.60	15.91	73.82	17.07	162.59
Percent of total	27.9	31.2	34.9	36.9	39.7	25.5	34.6	39.5	52.8
Household-commercial	587.23	65.85	56.88	90.99	37.44	23.01	72.46	18.18	74.94
Percent of total	33.7	36.0	33.2	35.8	28.8	36.9	34.0	42.1	24.3
Nonenergy	101.20	7.24	8.13	13.58	13.60	10.44	10.01	0.75	19.50
Percent of total	5.8	4.0	4.7	5.3	10.7	16.7	4.7	1.7	6.3
Nonelectricity transformation losses	-20.85	3.53	3.81	-3.95	-2.66	0.34	6.05	-0.80	-15.01
Percent of total	-1.2	1.9	2.2	-1.6	-2.0	0.5	2.8	-1.9	-4.9

Note: Percentages may not add to 100 due to rounding. Minus sign = gains.

TABLE B-5 *Per Capita Energy Consumption by Sector, 1972*
(tons oil equivalent per capita)

Sectors	U.S.	Canada	France	W. Germany	Italy	Netherlands	U.K.	Sweden	Japan
Total consumption	8.354	8.379	3.312	4.116	2.389	4.678	3.814	5.312	2.905
Transformation losses	1.410	1.898	.585	.678	.348	.602	.863	1.336	.504
Energy industry	.759	.606	.237	.297	.125	.367	.274	.166	.166
Transport	1.844	1.444	.489	.528	.356	.493	.492	.604	.359
Industrial	1.745	1.832	.912	1.194	.736	.935	1.083	1.374	1.130
Household-commercial	2.111	2.268	.931	1.199	.574	1.494	.922	1.740	.562
Nonenergy use	.485	.331	.157	.220	.250	.783	.179	.092	.184

TABLE B-6 *Energy Consumption Relative to GDP, by Sector, 1972*
(tons oil equivalent per million dollars GDP)

Sectors	U.S.	Canada	France	W. Germany	Italy	Netherlands	U.K.	Sweden	Japan
Total consumption	1,480	1,772	795	1,031	915	1,272	1,121	1,062	849
Transformation losses	250	401	140	170	133	164	254	267	147
Energy industry	135	129	57	74	48	100	81	33	48
Transport	327	305	117	132	136	134	145	121	105
Industrial	309	388	219	299	282	254	318	275	330
Household-commercial	374	480	223	300	220	407	271	349	164
Nonenergy use	86	70	38	55	96	213	53	18	54

TABLE B-7 *Energy Consumption in the Energy Sector, 1972*
(absolute figures in million tons oil equivalent)

Components of sector	U.S.	Canada	France	W. Germany	Italy	Netherlands	U.K.	Sweden	Japan
Total, energy sector	158.55	13.24	12.27	18.30	6.79	4.89	15.32	1.35	17.57
Hard coal and lignite mines	1.95	0.15	0.66	2.52	.01	0.07	1.23	0.0	0.86
Percent of total	1.2	1.1	5.4	13.8	0.1	1.4	8.0	0.0	4.9
Natural gas extraction	52.93	7.03	0.0	0.46	0.15	0.57	0.0	0.0	0.20
Percent of total	33.4	53.1	0.0	2.5	2.2	11.7	0.0	0.0	1.1
Crude petrol. extraction	0.0	0.0	0.07	0.0	0.0	0.0	0.0	0.0	0.0
Percent of total	0.0	0.0	0.6	0.0	0.0	0.0	0.0	0.0	0.0
Coke ovens	6.68	0.96	1.47	2.80	0.71	0.18	1.60	0.22	1.54
Percent of total	4.2	7.3	12.0	15.3	10.5	3.7	10.4	16.3	8.8
Gas works	neg.	0.0	0.08	0.21	0.02	0.0	0.28	0.02	0.58
Percent of total	neg.	0.0	0.7	1.1	0.3	0.0	1.8	1.5	3.3
Petroleum refineries	62.09	2.51	7.49	8.52	3.87	3.45	7.08	0.32	9.69
Percent of total	39.2	19.0	61.0	46.6	57.0	70.6	46.2	23.7	55.2
Electric plants	9.73	0.52	0.65	1.51	0.48	0.20	1.58	0.14	1.15
Percent of total	6.1	3.9	5.3	8.3	7.1	4.1	10.3	10.4	6.5
Transmission losses	25.16	2.07	1.84	2.07	1.38	0.43	3.46	0.65	3.32
Percent of total	15.9	15.6	15.0	11.3	20.3	8.8	22.6	48.1	18.9
Gas	9.89	0.46	0.92	0.86	0.44	0.21	1.84	0.05	0.96
Electricity	15.27	1.61	0.92	1.21	0.94	0.22	1.62	0.60	2.36
Pumped storage	0.0	0.0	0.02	0.21	0.17	0.0	0.10	0.00	0.23
Percent of total	0.0	0.0	0.2	1.1	2.5	0.0	0.7	0.00	1.3

Note: neg. = negligible.

TABLE B-8 *Per Capita Energy Consumption in the Energy Sector, 1972*
(tons oil equivalent per capita)

Components of sector	U.S.	Canada	France	W. Germany	Italy	Netherlands	U.K.	Sweden	Japan
Total, energy sector	.759	.606	.237	.297	.125	.367	.274	.166	.166
Hard coal, lignite, patent fuel, briquettes	.009	.007	.013	.041	neg.	.005	.022	.000	.008
Natural gas extraction	.253	.322	.000	.007	.003	.043	.000	.000	.002
Crude petrol. extraction	.000	.000	.001	.000	.000	.000	.000	.000	.000
Coke ovens	.032	.044	.028	.045	.013	.014	.029	.027	.015
Gas works	neg.	.000	.002	.003	neg.	.000	.005	.002	.005
Petrol. refineries	.297	.115	.145	.138	.071	.259	.127	.039	.091
Electric plants	.047	.024	.013	.024	.009	.015	.028	.017	.011
Transmission losses	.120	.095	.036	.034	.025	.032	.062	.080	.031
Gas	.047	.021	.018	.014	.008	.016	.033	.006	.009
Electricity	.073	.074	.018	.020	.017	.017	.029	.074	.022
Pumped storage	.000	.000	neg.	.003	.003	.000	.002	neg.	.002

Note: neg. = negligible.

TABLE B-9 *Energy Consumption in the Energy Sector Relative to GDP, 1972*
(tons oil equivalent per million dollars GDP)

Components of sector	U.S.	Canada	France	W. Germany	Italy	Netherlands	U.K.	Sweden	Japan
Total, energy sector	134.5	128.1	56.9	74.4	47.8	100.0	80.6	33.2	48.4
Hard coal and lignite	1.7	1.5	3.1	10.2	0.1	1.4	6.5	0.0	2.4
Natural gas extraction	44.9	68.0	0.0	1.9	1.1	11.6	0.0	0.0	0.6
Crude petrol. extraction	0.0	0.0	0.3	00.0	0.0	0.0	0.0	0.0	0.0
Coke ovens	5.7	9.3	6.8	11.4	5.0	3.7	8.4	5.4	4.2
Gas works	neg.	0.0	0.4	0.9	0.1	0.0	1.5	0.5	1.6
Petrol. refineries	52.7	24.3	34.8	34.6	27.2	70.4	37.3	7.9	26.7
Electric plants	8.3	5.0	3.0	6.1	3.4	4.1	8.3	3.4	3.2
Transmission losses	21.3	20.0	8.5	8.4	9.7	8.8	18.2	16.0	9.2
Gas	8.4	4.5	4.3	3.5	3.1	4.3	9.7	1.2	2.6
Electricity	13.0	15.6	4.3	4.9	6.6	4.5	8.5	14.8	6.5
Pumped storage	00.0	0.0	neg.	0.9	1.2	0.0	0.5	neg.	0.6

Note: neg. = negligible.

TABLE B-10 Energy Consumption in the Transport Sector, 1972
(absolute figures in million tons oil equivalent)

Components of sector	U.S.	Canada	France	W. Germany	Italy	Netherlands	U.K.	Sweden	Japan
Total, transport sector	385.11	31.55	25.30	32.59	19.36	6.57	27.49	4.97	38.07
Air	49.26	2.21	1.43	1.00	1.51	0.10	1.58	0.20	1.60
Percent of total	12.8	7.0	5.7	3.1	7.8	1.5	5.7	4.1	4.2
Road	319.45	26.97	21.87	28.36	16.70	5.19	23.53	4.11	28.68
Percent of total	83.0	85.5	86.4	87.0	86.3	79.0	85.6	83.7	75.3
Rail	16.40	2.34	1.29	2.25	0.65	0.15	1.43	0.26	3.55
Percent of total	4.3	7.4	5.1	6.9	3.4	2.3	5.2	5.3	9.3
Internal and coastal navigation	n.a.	0.04	0.71	0.98	0.50	1.13	0.95	0.34	4.24
Percent of total	n.a.	0.1	2.8	3.0	2.6	17.2	3.5	6.9	11.1

Note: n.a. = not available.

TABLE B-11 Per Capita Energy Consumption in the Transport Sector, 1972
(tons oil equivalent per capita)

Components of sector	U.S.	Canada	France	W. Germany	Italy	Netherlands	U.K.	Sweden	Japan
Transport sector	1.844	1.444	0.489	0.528	0.356	0.493	0.492	0.604	0.359
Air	0.236	0.101	0.028	0.016	0.028	0.008	0.028	0.025	0.015
Road	1.530	1.234	0.423	0.460	0.307	0.389	0.421	0.506	0.271
Rail	0.079	0.107	0.025	0.036	0.012	0.011	0.026	0.032	0.034
Internal and coastal navigation	n.a.	0.002	0.014	0.016	0.009	0.085	0.017	0.042	0.040

Note: n.a. = not available.

TABLE B-12 Energy Consumption in the Transport Sector Relative to GDP, 1972
(tons oil equivalent per million dollars GDP)

Components of sector	U.S.	Canada	France	W. Germany	Italy	Netherlands	U.K.	Sweden	Japan
Transport sector	327	305	117	132	136	134	145	121	105
Air	42	21	7	4	11	2	8	5	4
Road	271	261	101	115	117	106	124	101	79
Rail	14	23	6	9	5	3	8	6	10
Internal and coastal navigation	n.a.	0	3	4	4	23	5	8	12

Note: n.a. = not available.

TABLE B-13 Energy Consumption in the Industrial Sector, 1972
(absolute figures in million tons oil equivalent)

Components of sector	U.S.	Canada	France	W. Germany	Italy	Netherlands	U.K.	Sweden	Japan
Total, industrial sector	364.40	40.04	47.17	73.62	40.05	12.47	60.52	11.17	119.71
Iron and steel	68.87	4.63	12.86	24.38	7.95	2.76	15.55	2.42	45.87
Percent of total	18.9	11.6	27.3	33.1	19.9	22.1	25.7	21.7	38.3
Chemical		0.98	6.24	12.44		6.28	5.82	0.57	12.33
Percent of total		2.4	13.2	16.9		50.4	9.6	5.1	10.3
Other	295.52[a]	34.43	28.07	36.81	32.12[a]	3.44	39.16	8.17	61.51
Percent of total	81.1[a]	86.0	59.5	50.0	80.2[a]	27.6	64.7	73.1	51.4

[a] The figures for "Other" include "Chemical" in these cases.

TABLE B-14 *Per Capita Energy Consumption in the Industrial Sector, 1972*
(tons oil equivalent per capita)

Components of sector	U.S.	Canada	France	W. Germany	Italy	Netherlands	U.K.	Sweden	Japan
Total, industrial sector	1.745	1.832	.912	1.194	.736	.935	1.083	1.374	1.130
Iron and steel	.330	.212	.249	.395	.146	.207	.278	.298	.433
Chemical		.045	.121	.202		.471	.104	.070	.116
Other	1.415[a]	1.576	.543	.597	.590[a]	.258	.701	1.005	.580

[a] The figures for "Other" include "Chemical" in these cases.

TABLE B-15 *Energy Consumption in the Industrial Sector Relative to GDP, 1972*
(tons oil equivalent per million dollars GDP)

Components of sector	U.S.	Canada	France	W. Germany	Italy	Netherlands	U.K.	Sweden	Japan
Total, industrial sector	309	388	219	299	282	254	318	275	330
Iron and steel	58	45	60	99	56	56	82	60	126
Chemical		9	29	51		128	31	14	34
Other	251[a]	333	130	150	226[a]	70	206	201	170

[a] The figures for "Other" include "Chemical" in these cases.

TABLE B-16 Energy Consumption in the Household-commercial Sector, 1972
(absolute figures in million tons oil equivalent)

Components of sector	U.S.	Canada	France	W. Germany	Italy	Netherlands	U.K.	Sweden	Japan
Total, household-commercial sector	440.85	49.56	48.13	73.92	31.25	19.96	51.53	14.15	59.51
Household	256.9	29.9	27.2	40.3	19.0	12.0	34.3	9.4	30.7
Percent of total	58.3	60.3	56.5	54.5	60.8	60.1	66.6	66.4	51.6
Commercial (incl. misc.)	155.0	14.6	18.1	31.7	10.3	5.9	15.2	4.3	26.1
Percent of total	35.2	29.5	37.6	42.9	33.0	29.6	29.5	30.4	43.9
Agricultural	29.0	5.0	2.8	1.9	2.0	2.1	2.0	0.5	2.7
Percent of total	6.6	10.1	5.8	2.5	6.4	10.5	3.9	3.5	4.5

TABLE B-17 Per Capita Energy Consumption in the Household-commercial Sector, 1972

(tons oil equivalent per capita)

Categories of sector	U.S.	Canada	France	W. Germany	Italy	Netherlands	U.K.	Sweden	Japan
Total, household-commercial sector	2.11	2.27	0.93	1.20	0.57	1.50	0.92	1.74	0.56
Household	1.23	1.37	0.53	0.65	0.35	0.90	0.61	1.16	0.29
Commercial (incl. misc.)	0.74	0.67	0.35	0.51	0.19	0.44	0.27	0.53	0.25
Agricultural	0.14	0.23	0.05	0.03	0.04	0.16	0.04	0.06	0.03

TABLE B-18 Energy Consumption in the Household-commercial Sector Relative to GDP, 1972

(tons oil equivalent per million dollars GDP)

Categories of sector	U.S.	Canada	France	W. Germany	Italy	Netherlands	U.K.	Sweden	Japan
Total, household-commercial sector	374	480	223	300	220	407	271	348	164
Household	218	289	126	164	134	245	180	231	85
Commercial (incl. misc.)	132	141	84	129	72	120	80	106	72
Agricultural	25	48	13	8	14	43	11	12	7

TABLE B-19 Energy Consumption for Nonenergy Uses, 1972

Consumption	U.S.	Canada	France	W. Germany	Italy	Netherlands	U.K.	Sweden	Japan
Nonenergy uses (million tons oil equiv.)	101.20	7.24	8.13	13.58	13.60	10.44	10.01	0.75	19.50
Percentage of total energy consumption	5.8	4.0	4.7	5.3	10.5	16.7	4.7	1.7	6.3
Nonenergy uses (tons oil equiv. per capita)	0.485	0.331	0.157	0.220	0.250	0.783	0.179	0.092	0.184
Nonenergy uses (tons oil equiv. per $ million GDP)	86	70	38	55	96	213	53	18	54

TABLE B-20 Total Energy Consumption by Fossil Fuels and Primary Electricity, 1972
(absolute figures in million tons oil equivalent)

Consumption	U.S.	Canada	France	W. Germany	Italy	Netherlands	U.K.	Sweden	Japan
Total energy consumption	1,744.65	183.09	171.23	253.85	130.01	62.36	213.10	43.19	307.82
Fossil fuels									
Solids	307.81	17.35	30.39	85.43	8.41	3.35	73.14	1.51	50.58
Percent of total	17.6	9.5	17.7	33.7	6.5	5.4	34.3	3.5	16.4
Petroleum and products	802.07	85.64	113.96	139.94	97.39	32.72	107.96	27.88	230.94
Percent of total	46.0	46.8	66.6	55.1	74.9	52.5	50.7	64.6	75.0
Natural gas	551.80	34.66	11.75	21.88	12.86	26.34	23.72	0.19	2.50
Percent of total	31.6	18.9	6.9	8.6	9.9	42.2	11.1	0.4	0.8
Primary electricity	82.97	45.44	15.13	6.60	11.35	0.05	8.28	13.61	23.80
Percent of total	4.8	24.8	8.8	2.6	8.7	neg.	3.9	31.5	7.7

Note: neg. = negligible.

TABLE B-21 *Total Final Consumption by Fossil Fuels and Electricity, 1972*
(absolute figures in tons oil equivalent)

	U.S.	Canada	France	W. Germany	Italy	Netherlands	U.K.	Sweden	Japan
Consumption by final consuming sectors	1,450.12	141.62	141.01	212.02	111.05	54.33	164.86	32.33	254.36
Fossil fuels	1,280.30	121.19	126.79	187.37	99.42	50.19	142.14	26.05	217.49
percent	88.3	85.6	89.9	88.4	89.5	92.4	86.2	80.6	85.5
Electricity	169.82	20.43	14.22	24.65	11.64	4.14	22.72	6.28	36.87
percent	11.7	14.4	10.1	11.6	10.5	7.6	13.8	19.4	14.5
Energy sector	158.55	13.24	12.27	18.30	6.79	4.89	15.32	1.35	17.57
Fossil fuels	133.54	10.80	10.09	14.34	5.02	4.32	11.26	0.59	13.18
percent	84.2	81.6	82.2	78.4	73.9	88.3	73.5	43.7	75.0
Electricity	25.01	2.44	2.18	3.96	1.77	0.57	4.06	0.76	4.39
percent	15.8	18.4	17.8	21.6	26.1	11.7	26.5	56.3	25.0
Transport sector	385.11	31.55	25.30	32.59	19.36	6.57	27.49	4.91	38.07
Fossil fuels	384.73	31.51	24.78	31.85	19.04	6.49	27.26	4.74	36.99
percent	99.9	99.9	97.9	97.7	98.6	98.8	99.2	96.5	97.6
Electricity	0.38	0.05	0.52	0.74	0.32	0.08	0.23	0.17	1.08
percent	0.1	0.1	2.1	2.3	1.4	1.2	0.8	3.5	2.4
Industrial sector	364.40	40.04	47.17	73.62	40.05	12.47	60.52	11.17	119.71
Fossil fuels	298.79	30.85	40.37	62.87	33.82	10.62	53.36	7.99	96.62
percent	82.0	77.0	85.6	85.4	84.4	85.2	88.2	71.5	80.7
Electricity	65.61	9.19	6.80	10.75	6.22	1.85	7.16	3.18	23.09
percent	18.0	23.0	14.4	14.6	15.6	14.8	11.8	28.5	19.3
Household-commercial sector	440.85	49.56	48.13	73.92	31.25	19.96	51.53	14.15	59.51
Fossil fuels	362.03	40.79	43.42	64.73	27.92	18.32	40.26	11.98	51.20
percent	82.1	82.3	90.2	87.6	89.3	91.8	78.1	84.7	86.0
Electricity	78.82	8.77	4.71	9.19	3.33	1.64	11.27	2.17	8.31
percent	17.9	17.7	9.8	12.4	10.7	8.2	21.9	15.3	14.0
Nonenergy sector	101.20	7.24	8.13	13.58	13.60	10.44	10.01	0.75	19.50
Fossil fuels	101.20	7.24	8.13	13.58	13.60	10.44	10.01	0.75	19.50
percent	100.0	100.0	100.0	100.0	100.0	100.0	100.0	100.0	100.0

Estimating Procedures Used in Analysis of the Household-commercial Sector

Estimating Energy Consumed in Agricultural, Space Conditioning, Household, and Commercial Uses

Developing estimates of energy consumed in the household-commercial sector[1] was a step-by-step process. The point of departure for these estimates was "total energy consumption."

Agriculture

The first step was to estimate consumption by the agricultural sector. Much of this data was given by the Organisation for Economic Co-operation and Development (OECD) in their energy statistics, but for some countries the coverage was incomplete and had to be supplemented by data available from national sources. It should be noted that energy consumed in agriculture includes energy consumption in agricultural transport.

Space Conditioning

The energy consumption of the agricultural sector, derived as above, was then subtracted from the total household-commercial sector to yield a subtotal that was partitioned first of all into space conditioning and non-space conditioning and second, into household and commercial.

For some countries—United States, United Kingdom, and Netherlands—firm data on the consumption of fuels in space conditioning were available from national sources. For other countries it was necessary to make estimates based on the only available breakdown of energy consumption

[1] For convenience in our study we have called this sector "household-commercial." The more accurate term, however, is the OECD term "other sector," that is, the sector that includes all energy consumption not included within more clearly defined sectors.

in these sectors, type of fuel. At first sight this may seem to be an inexact basis for estimation. But many countries—France, West Germany, Italy, Sweden, and Japan—rely heavily on petroleum products, of which home heating oil was separately distinguished. This clearly provided a very reasonable basis for estimation. For the other countries—United States, Canada, Netherlands, and the United Kingdom—the pattern of fuel consumption in this sector is more complex. Fortunately these were the countries for which national estimates were available so that it was not strictly necessary to make estimates based on the pattern of fuel consumption. Nonetheless, in order to test the accuracy of the approach for the countries for which data were not available, estimates were prepared. These were then checked against the national estimates and found in all cases to be very close. Such proximity gives grounds for confidence in the estimates prepared for the countries which have no independent, separate consumption data.

The following assumptions were made with regard to estimating energy use in space conditioning on the basis of fuel consumption:

1. All coal is used for heating rather than nonheating purposes. Note that the small amount of coal used in most countries means that the estimates would not have been much different even if a substantial percentage of the total had been assigned to nonheating use. France, Germany, and United Kingdom are substantial coal users so that this assumption may overestimate their heating consumption, though independent data for the United Kingdom suggest that the original assumption is reasonable.

2. In general it is assumed that electricity is used mainly for nonspace conditioning uses. There are three modifications to this assumption. First, the United States consumes substantial amounts of electricity in space conditioning, particularly for air conditioning. (It was assumed that the air conditioning consumption of the other countries is negligible.) Second, the United Kingdom has exceptionally large electricity consumption in this sector, part of which is used for space heating in the form of small heaters. Third, of the other countries, Sweden is known to have a certain amount of electric heating (7 percent of the houses are centrally heated by electricity). Adjustments were made for all three countries, but for the others, electric heating and cooling were assumed to be negligible.

3. Fortunately, petroleum products for heating purposes are distinguished separately for each country. Given the importance of petroleum products in the energy consumption pattern of the household and

commercial sectors, this gives an excellent indication of the amount of fuel used for heating. The only complication in this sector occurred in those countries, Japan in particular, where significant quantities of kerosine (which can be used for cooking and lighting as well as heating) are consumed. In the case of Japan almost all kerosine was allocated to heating rather than nonheating use on the basis that any other solution would have left Japan's consumption for heating purposes unreasonably low.

4. This leaves gas. For those countries with significant supplies of natural gas, national statistics were available. For countries using only town gas, it was assumed that most went to nonheating purposes.

Estimates of energy used in space conditioning as opposed to nonspace conditioning arrived at by this procedure are given in table C-1. Space conditioning is the major functional end use of energy in this sector for all countries, accounting typically for 70 to 80 percent. For all except the United States, electricity accounts for 10 percent or less of heating and cooling fuels, although, as noted above, the estimating procedure used would tend to underestimate the part of electricity.

The relatively heavy use of electricity in the United States reflects the widespread use of air conditioning. In 1972 air conditioning was little used in Europe, Japan, or even Canada (which in so many respects resembles the U.S. consumption pattern). For this reason, in the analysis of the household category, the consumption of energy in air conditioning was extracted from total space conditioning in the case of the United States and assumed to be negligible for all other countries, thus permitting a direct comparison between countries of fuel and power consumption for heating only.

The estimates of energy consumed in space conditioning were limited in some respects. For example, water heating, which is a major user of energy (accounting for about 4 percent of total energy consumption in the United States and for 12 percent of total energy consumption in the household-commercial sector) was, in principle, included in nonspace conditioning uses. But for some European countries, part of the energy used in heating, particularly in household central heating systems based on hot water radiators, may in practice be used to heat water. The bias of such ambiguity would be to overestimate countries' energy consumption in heating relative to the United States.

In addition, the data do not include either noncommercial fuel or waste heat sold for heating purposes by generating stations. The exclusion of noncommercial fuel was assumed not to be of major significance because all of our countries are highly industrialized urban economies. Such

TABLE C-1 *Energy Consumption for Agricultural, Household, Commercial, Space Conditioning, and Non-space Conditioning Uses, 1972*

Uses	U.S.	Canada	France	W. Germany	Italy	Netherlands	U.K.	Sweden	Japan
Energy consumption, household-commercial sector [a] (million tons oil equiv.)	440.9	49.6	48.1	73.9	31.3	20.0	51.5	14.2	59.5
Agriculture	29.0	5.0	2.8	1.9	2.0	2.1	2.0	0.5	2.7
As percent of sector	6.6	10.1	5.8	2.6	6.4	10.5	3.9	3.5	4.5
Household	256.9	29.9	27.2	40.3	19.0	12.0	34.3	9.4	30.7
As percent of sector	58.3	60.3	56.5	54.5	60.7	60.0	66.6	60.2	51.6
Commercial	155.0	14.7	18.1	31.7	10.3	5.9	15.2	4.3	26.1
As percent of sector	35.2	29.6	37.6	42.9	32.9	29.5	29.5	30.3	43.9
Space conditioning	322.0	37.7	38.7	58.9	24.0	12.8	36.0	12.4	39.0
As percent of sector	73.0	76.0	80.4	79.7	76.7	64.0	69.9	87.3	65.5
Non-space conditioning	89.9	7.0	6.7	13.1	5.3	5.1	13.5	1.2	17.8
As percent of sector	20.4	14.1	13.9	17.7	16.9	25.5	26.2	9.2	29.9

Note: Percentages may not add to 100 due to rounding.

Source: RFF estimates.

[a] Termed "other sector" in OECD data.

fuels—particularly wood and charcoal—may nonetheless provide a significant supplement to commercial fuels in countries such as Japan and Italy, and in the large number of vacation houses in Sweden. Because district heating is believed to account for 15 percent of all space heating in Sweden and for lesser but still significant amounts in Germany and the United Kingdom, the exclusion of waste heat and district heating from this sector will similarly underestimate the space heating energy consumption of these countries.

Finally, the estimates and the analysis based on the estimates obscure important regional differences within countries. Patterns of energy consumption in space heating, for example, probably vary more within the United States than they do between the United States and other countries. There is a similar situation in Italy, where the climatic spread between north and south is equally large.

Household and Commercial Uses

The total consumption for space conditioning and non-space conditioning was then repartitioned into household and commercial uses. The main purpose was to establish household uses, commercial uses being taken as the residual. Again, the first breakdown was based primarily on fuel consumption. The following assumptions were made:

1. coal—entirely allocated to residential uses
2. gas and electricity—estimates made on basis of data given in Economic Commission for Europe bulletins for these fuels
3. petroleum products—as a first estimate, 60 percent was allocated to the household rather than the commercial category.

The resulting estimates were then compared with national estimates and adjusted where necessary. A check was made to ensure internal consistency—that is, that the heating portion of household use plus the heating portion of commercial use equaled total heating use. In practice this requirement, coupled with some rough and ready criteria regarding credibility of various consumption categories, imposed quite narrow limitations on the distribution of total supplies within the various categories. The estimates are given in table C-1.

Estimating Degree Days

To assess the influence of climate on levels of heating fuel consumption it was necessary to make an estimate of degree days for our countries (see table C-2). A certain amount of data on degree days already existed. The problem was therefore (1) to estimate degree days for those

TABLE C-2 *Degree Day Data, 1972*

Degree days and consumption	U.S.	Canada	France	W. Germany	Italy	Netherlands	U.K.	Sweden	Japan
Centigrade degree days based on 16°C (60.8°F) threshold	2,030	3,700	2,200	2,600	1,700	2,725	2,200	3,800	2,000
Centigrade degree days based on 18°C (65°F) threshold	2,700	4,200	2,700	3,500	2,100	3,290	2,840	4,300	2,400
Per capita consumption of heating fuel in household and commercial uses (tons oil equiv.)	1.44	1.73	0.75	0.96	0.44	0.96	0.64	1.53	0.37
Consumption per capita per centigrade degree day (16° threshold) (kilograms oil equiv.)	0.71	0.47	0.34	0.37	0.33	0.35	0.29	0.40	0.19

Source: Degree-day data are RFF estimates based on national sources and on U.S. Department of Agriculture, *Weekly Weather and Crop Bulletin,* vol. 61, no. 5 (November 1974).

countries for which data were not available and (2) to adjust degree-day data where necessary to a standardized threshold basis. In Europe the threshold temperature from which degree days are counted is 60.8° Fahrenheit, or 16° Celsius. Data based on this were available for France, West Germany, the Netherlands, and Sweden. Consequently, we had to prepare estimates for the United States, Canada, Italy, the United Kingdom, and Japan. To do this, we first plotted the coolest monthly temperatures for both the capital city and the country against degree days for those countries where degree-day information was available. Readings of degree days were then taken for the missing countries based on their coolest monthly temperatures. In most countries, capital and country average coincided. For Italy, which has a larger climatic difference than many other countries, the country rather than the capital was taken as more representative.

A problem existed for the United States and Canada because their degree days (available only from national sources) were based on a 65° Fahrenheit, or 18° Celsius, threshold. It was therefore necessary to prepare estimates for these countries on a 60.8° Fahrenheit, or 16° Celsius, basis in order to facilitate comparison with the European countries. This was done for the United States and for Canada by deducting the average monthly temperatures in the main metropolitan areas from the 60.8°F threshold and weighting the results by population to arrive at national averages.

Table C-2 gives degree days for all our countries calculated on two bases, the European threshold standard of 60.8°F and the North American threshold of 65°F. Under both systems Canada and Sweden— among the heavy consumption countries it will be remembered—have a large number of degree days, and Italy, the smallest. The others are all fairly closely grouped, with the United States among the lowest. It already becomes clear that degree days, that is, climatic differences, account for some of the intercountry variations between heating fuel consumption, but not all, and particularly not the high U.S. consumption.

Assessing Housing Characteristics

Table 4-5 in the text of this study gives data on housing characteristics of our nine countries. The problem here was to assess the effect of different types of housing in the various countries on their consumption of heating fuel. Given the patchiness of the data it was considered best to do this by making a composite index of housing conditions, particularly as they affect the consumption of heating fuel.

The two characteristics incorporated in the index are size of house and percentage of single family, rather than multiple family, homes. These two items were selected because data were available on them and also because they were assumed to be the housing characteristics most relevant to consumption of heating fuel. Insulation standards are also of importance. However, as most countries, apart from Sweden and Canada, were assumed to have very similar standards of insulation, these were not incorporated into the index.

The first step was to construct an index of housing size, based on the information in table 4-5, supplemented where necessary by estimates. This index was then modified to take account of the fact that consumption of fuel for heating tends to increase less than proportionately with house size. It was assumed in fact that a doubling in house size was accompanied by a 50 percent increase in heating fuel.[2]

A second step was to construct an index based on the percentage shares of single family homes in the total housing stock. Their effect on heating fuel consumption was calculated by weighting these shares by the assumed 50 percent greater heating needs of the single family houses of a given size compared with the multiple family houses.[3]

These two indexes were then multiplied to give a composite index (see table 4-5). This index sets U.S. housing characteristics at 100. The values of other countries are lower, reflecting a smaller average size of house, a smaller proportion of single family homes, or both. Most countries have an index value of about 70, except for United Kingdom (with a high percentage of single family homes) and Sweden (large houses), both at about 80. The composite housing index was less useful for Japan than for other countries because Japan's housing differs widely.

This index must be used with caution. It is based on housing data which are limited in coverage and frequently not comparable between countries. In several cases estimates based on judgement were made. There was also some overlap between the two indexes that went into the composite index. The index of house size will have already incorporated some of the mix between single and multifamily units insofar as single family units tend to be larger than multifamily units. This duplication would tend to lower the values of other countries relative to the United States, though in practice not by very much. This index was based on average size of house rather than the total housing stock of the various

[2] Nigel Lucas, Malcolm Newton, Ted Nicklin, and Tony Wickens, *Energy: Waste Not Want Not* (Tyneside, England, Environmental Concern, August 1974).

[3] A. Doernberg, *Comparative Analysis of Energy Use in Sweden and the United States* (Upton, N.Y., Brookhaven National Laboratory, 1975).

countries. Though comprehensive data were not available, it appeared that basing the index on housing stock rather than average size would tend to increase the values.

Insulation practices were assumed to be very similar in most countries except for Sweden and Canada. (For example, U values for walls range between 0.20 and 0.30 Btu per square foot in United Kingdom, United States, and Netherlands, compared with 0.12 in Sweden.) But allowance for the superior insulation standards of Sweden and Canada has not been made in the composite index.

It will be clear from the above that any attempt to attribute total differences in energy consumption for heating purposes to housing stock must be regarded as conjectural. The typically larger single family house is responsible for some of the difference between countries, but the exact share is more difficult to pin down.

Measuring the Efficiency of Heating Equipment

As illustrated in table 4-7 in the text of this study, the efficiency of heating equipment was measured as the amount of "useful" energy which emerges from a given energy input. This depends on the type of heating equipment and the fuel used. Efficiencies assigned to different systems depend on whether design or actual efficiencies were used; whether hot air or hot water systems or central or local heating systems predominated; and whether heat losses were incorporated into the calculation of electricity efficiency. Here, actual rather than design efficiencies were used, and heat losses were not included in electricity efficiencies. Some adjustment was made for the prevalence of central rather than local heating systems. The resulting efficiencies are given in table 4-7. These efficiencies, multiplied by the fuels used in each country, gives an amount of "useful" energy obtained from the total consumption. Three groups of countries emerge, the United States, Canada, and Sweden with relatively high efficiencies (about 70 percent) owing to a combination of a high proportion of central heating systems and very little use of coal. Most other countries have efficiencies of about 60 to 65 percent owing to less central heating and greater use of coal. If the U.S. electricity use had been counted at a 30-percent rather than a 94-percent efficiency (to take account of heat losses), the U.S. efficiency would also have fallen to about 60 percent. The exceptional country is the United Kingdom, whose useful energy is only one-half of total input because of the prevalence of local heating with coal.

Derivation of Data on Energy Consumption and Passenger-miles of Travel, by Passenger Transport Modes

General Notes

BASIC TO THE PURPOSE of chapter 5 was the necessity of developing for each of our nine countries estimates of passenger transport energy use and of passenger-miles traveled for each of the principal passenger travel modes—cars, buses, rail, and air. Only with such data in hand can one undertake even a rudimentary review of the extent to which intercountry differences in overall energy/GDP ratios are influenced by the volume and modal patterns of passenger transportation activity and by the energy intensity of each transport mode.

Unfortunately, the statistical building blocks needed for this undertaking were found to be in deplorable shape. There is no assembly of standardized, comparative international data on energy consumed in passenger transportation or on passenger-miles generated. (Some international *collections* of passenger-mile data—a different proposition—do exist, and we made all possible use of these. But these were quite deficient in matters of comparability and completeness.) Frequently, even recourse to selected national data sources—when not leading to gaps and dead ends—exposed significant inconsistencies.

Nevertheless, even given these limitations, the importance of passenger transport as a factor in international energy/GDP variability prompted us to try to construct the best set possible of country-by-country figures on energy consumption and passenger mileage. The data contained in table 5-3 are the results of this effort. In one way or another, these figures—which, if far from ideal, we deem at least serviceable for our needs—enter recurrently into the analysis in chapter 5. The following paragraphs detail the scope and character of the data, and draw attention to some of the more vexing problems we encountered.

1. For most countries (except for the United States and Canada) statistics on energy consumed in passenger transportation could only be obtained independently of corresponding data on passenger-miles traveled. This meant that the quotient of these two indicators—the critically important energy intensity measure—might be distorted to a greater or lesser degree by errors in numerator (energy consumed) or denominator (passenger-miles). Particularly where the absolute size of numerator or denominator was relatively small (as in the case of air travel) this tendency for distorted energy intensities appeared to worsen. Indeed, in a few cases, where comparative energy intensities among the countries seemed to us to gyrate outside sensible upper or lower limits, we used our judgment to impose reasonable bounds—never, however, without first ensuring that such an adjustment would not significantly alter the weighted energy intensity of a country's aggregated modal components. And, of course, where (for example, in tables 5-3 and 5-4) we draw comparisons between the United States, on the one hand, and a combined group of other countries, on the other, imperfections in individual foreign country data can reasonably be expected to wash out.

2. In principle, the data refer solely to domestic travel. For the most part, we think that our figures conform to that guideline, although we were unable to resolve known statistical uncertainties completely. For example, fuel consumption data are more likely to be limited to internal consumption, while associated passenger mileage may, to an unknown degree, include domestic as well as foreign travel. This may produce some distortion in energy intensities for given countries, although mutual offsets among countries will cushion the extent of such distortion. Thus, the portion of French fuel consumption ascribable to German motorists is paralleled, at least to an extent, by German fuel consumption accounted for by French motorists. Another element of uncertainty in passenger transport statistics relates to air transport in which for numerous countries, the statistics usually include (and are often dominated by) foreign travel. We dealt with this problem to the extent possible—for example, by using figures on air bunker fuels as the basis for removing the foreign component—but we lack confidence as to the accuracy of the results.

3. The data include most, but not all, passenger travel modes. Inland and coastal navigation are excluded. Other intended exclusions—we cannot guarantee invariable success at having kept them out—are: nonairline aviation; school buses; and a number of statistically minor personal travel modes: motorcycles, mopeds, bicycles, and three-wheeled convey-

ances. In principle, business use of passenger cars and travel by taxis are included in the automotive category. Once again, we cannot vouch for entirely consistent treatment across the nine countries. The rail component includes long-distance railways, subways, commuter lines, and streetcars. It might have been instructive to differentiate among these in the analysis, but there was not even a remote chance of being able to do this in a standardized fashion for the nine countries.

4. Some numbers had to be estimated in a most indirect manner. Two illustrations make this point vivid. In one example it was impossible to locate any estimate of energy consumed by Dutch passenger trains. However, we could obtain an estimate of all Dutch rail traffic (passenger plus freight) from the annual OECD *Statistics of Energy*.[1] Next, we noted an estimate of ton-miles (actually, ton-kilometers in the first instance) of freight generated by railways in the Netherlands. In a study conducted on behalf of the European Communities ("La consommation d'énergie des moyens de transport" in 1974[2]) we found an estimate of freight energy intensity (energy consumed per ton-mile) by means of which, given the earlier estimate of ton-miles, we were able to construct a crude approximation of total energy used by freight trains. Subtracting this from the total estimated energy consumption accounted for by Dutch rail traffic yields the estimate of energy consumption by the passenger portion of Dutch rail traffic shown in table 5-3.

Another example began with the fact that no passenger-mile figure for cars was available in France. (The passenger-mile data which we *were* able to locate for France applied only to national roads and excluded, among other places, Paris.) We therefore had to resort to the estimation of passenger-miles by dividing an independently and crudely derived energy intensity estimate into an equally rough figure on total passenger-car energy consumption. (The energy intensity figure was based on separate figures for towns of over and under 5,000 provided by the Agence pour les Economies d'Energie, Ministère de l'Industrie et de la Recherche; the total consumption figure appears in Comité Professionel du Pétrole, *Eléments Statistiques 1972*.)

5. The energy intensities (100,000 tons of oil equivalent per passenger mile) recorded in table 5-3 are, as noted earlier, the direct outgrowth of these calculations. However, the intensity figures shown obviously reflect a number of underlying "efficiency" factors, of which (for example, in the

---

1 See list of sources consulted, at the end of this appendix.
2 See list of sources consulted, at the end of this appendix.

case of automobiles) the rated physical properties (engine displacement or vehicle weight) of the nation's car stock are undoubtedly the most important, but certainly not the only contributing elements. Others are the load factor (occupancy rate) of a country's motor vehicle population; the relative shares of urban and highway driving; and speed, maintenance, and driving habits. In analyses of transportation energy usage for given countries, these different elements have sometimes been addressed explicitly. For example, one analysis for the United States suggested energy intensity for urban driving that is approximately 140 percent above that for intercity traffic—a difference which would be shaved to 40 percent if intercity passenger loads applied in urban areas.[3] In the present nine-country survey, such contributing factors cannot be separately identified even though where possible we express cautious inference. The point is that one might interpret intercountry intensity differences almost exclusively on the basis of one's awareness, as a visitor to other countries, of differences in average car size. More is clearly involved.

6. The final rendering of the consumption figure into tons of oil equivalent represents a conversion from a variety of underlying physical and calorific units of measure. Generally, the next-to-last step was to record consumption in British thermal units (Btus), and then, following the OECD conversion practice, express everything in tons of oil equivalent at the rate of 40 million Btus per ton. Prior adjustments, in turn, frequently involved going from physical units (for example, gallons, barrels, or cubic meters of automotive fuel) to calorific units. Where possible, we used conversion factors applicable to the particular identifiable fuels (for example, 5.2 million Btus per American barrel of gasoline or 5.8 million Btus per barrel of diesel oil). However, sometimes we were forced to strike rough averages when consumption was known to cover different fuel categories. In the light of other statistical approximations surrounding chapter 5, not much is likely to have been lost thereby. For the electric component of the rail mode, our conversion is based on the inherent thermal content of electricity (3,412 Btus per kilowatt hour [kWh]) rather than the corresponding power station fuel in-

---

[3] Eric Hirst, *Energy Intensiveness of Passenger and Freight Transport Modes 1950–1970:* Report ORNL-NSF-EP-44 (Oak Ridge, Tenn., Oak Ridge National Laboratory, 1973) pp. 27, 32. Another study—a two-country comparison of Sweden and the United States was likewise able to explore some of these specific elements more thoroughly. (See Lee Schipper and A. J. Lichtenberg, *Efficient Energy Use and Well-Being: The Swedish Example* (Berkeley, Calif., Lawrence Berkeley Laboratory, University of California, 1976).

put (say 10,000 Btus per kWh). This conforms to our practice in this study of generally distributing final energy consumption among use-sectors on an "energy-delivered" basis, while allocating major conversion losses (as in electric power generation) to the transformation sector.[4] Thus, if electric traction prevailed in one country while diesel locomotives dominated in another, then equal units of final consumption could, in fact, signify a three-to-one comparison at the stage of primary energy requirements.

Sources Consulted

The first seven items listed below provided multicountry coverage, information, or analysis; the remaining references were used largely for material on given countries, as indicated. However, even those—as in the case of the SRI entry under Germany and the Over item under the Netherlands—often provided comparative data on other countries. The reader will note the absence of any specific published source for Italy. To our knowledge, none exists. As a result, the Italian data in table 5-3 are particularly weak because they lack any kind of validation through cross-checks between multicountry and national sources.

Multicountry Sources

1. Instituut voor Wegtransportmiddelen, "La consommation d'énergie des moyens de transport," prepared for the European Communities, Brussels, 1974.

2. European Communities, *Transport Yearbook* (Brussels, European Communities, 1973).

3. European Conference of Ministers of Transport, *Twentieth Annual Report* (The Hague and Paris, 1973).

4. International Civil Aviation Organization, *Airline Traffic, Volume One, Digest of Statistics, 1968–1972* (Montreal, ICAO, 1973).

5. International Road Federation, *World Road Statistics 1969–73* (Lausanne, Switzerland, 1974).

6. Organisation for Economic Co-operation and Development, *Statistics of Energy, 1958–72* (Paris, OECD, 1973).

7. United Nations Economic Commission for Europe, *Annual Bulletin of Transport Statistics, 1972* (Geneva, ECE, 1973).

---

[4] The interindustry analysis of chapter 7 is, for reasons explained there, an exception to this practice.

Country Sources

8. *For United States:* Federal Energy Administration, *Project Independence and Energy Conservation: Transportation Sector,* Final Task Force Report (Washington, D.C., GPO, 1974).

9. *For Canada:* J. Lukasiewicz, *Oil and Transportation in Canada and the United States,* Document ERG 75-1 (Ottawa, Carleton University, 1975); Statistics Canada, *Detailed Energy Supply Demand in Canada,* catalogue 57-207 annual, various issues.

10. *For France:* Comité Professionel du Pétrole, *Eléments Statistiques 1972* (Paris, CPP, 1973); data obtained from Agence pour les Economies d'Energie (Paris, Ministry of Industry and Research).

11. *For West Germany: Energiebilanz der Bundesrepublik Deutschland für das Jahr 1972* (Frankfurt, Verlags Wirtschaftsgesellschaft der Elektrizitätswerke GMBH, 1973); Stanford Research Institute, *Comparison of Energy Consumption Between West Germany and the United States* (Menlo Park, Calif., SRI, 1975).

12. *For Netherlands:* J. A. Over, ed., *Energy Conservation: Ways and Means* (The Hague, Future Shape of Technology Foundation, 1974).

13. *For United Kingdom:* National Economic Development Office, *Energy Conservation in the United Kingdom* (London, HMSO, 1974); National Economic Development Office, *The Increased Cost of Energy: Implications for Industry* (London, HMSO, 1974).

14. *For Sweden:* A. Doernberg, *Comparative Analysis of Energy Use in Sweden and the United States* (Upton, N.Y., Brookhaven National Laboratory, 1975); Statens offentliga Utredningar, Industridepartementet, Energiprognosutredningen, *Energi 1985–2000,* 2 vols., document 1974:65 (Stockholm, 1974).

15. *For Japan:* Japan, *Statistical Yearbook 1973/74;* data supplied by the Ministry of Transportation and the Institute of Energy Economics, Tokyo.

Procedure for Analyzing Intercountry Differences in the Ratio of Energy Consumption for Passenger Transport to Gross Domestic Product

AN IMPORTANT REASON for developing the data assembled in table 5-3 was to permit an analytical means of showing the extent to which variability in overall energy/GDP ratios arises from the characteristics of the transport systems in various countries. For reasons explained at the beginning of chapter 5, our effort centered largely on passenger transport.

In this appendix, we describe the basic computations.

As is clear from chapter 2, the ratio of energy consumption in a specific sector of the economy to national GDP merely provides an arithmetic description of the extent to which intercountry variability in overall energy/GDP ratios arises from difference in the relative proportions of sectoral energy use. For example, as the tabulation below shows, about 38 percent of the difference in the overall energy/GDP ratios between the United States and United Kingdom is attributable to differences in the relative importance of passenger transportation in total energy use:

	U.S.	U.K.	Difference
1. Ratio of energy consumption to GDP (tons oil equiv. per $ million GDP)	1480	1121	359
2. Ratio of energy consumption in passenger transport to GDP (tons oil equiv. per $ million GDP)	217	82	135
3. Line 2 as percent of line 1	15	7	38

In other words, we can point to passenger transport as "a big reason" for Britain's lower energy/GDP ratio. What the comparison does not do is to tell us why energy for passenger transport occupies a comparatively subsidiary role in Britain's energy picture. Specifically, is it because of

252

structural (mix) questions—that is, due to the fact that passenger transportation activity (measured by the relationship of passenger miles traveled to GDP) or the transportation modal mix (the relative importance of cars, buses, rail, and aircraft, with their different energy intensities) are appreciably different between the two countries? Or does it arise from differing energy intensities—energy consumed per passenger mile —for given modes? Or does the explanation stem from some combination of those factors?

In an effort to track these various possibilities, we developed the computational scheme illustrated in tables E-1 and E-2. In these, United States and United Kingdom are compared. Seven such additional sets of comparisons—between the United States, on the one hand, and each of the other countries, on the other—were performed. Chapter 3 itself includes a summary tabulation of all eight binary comparisons as well as the results of a comparison between the United States and the six-country Western European group as a whole.

In table E-1 we show the computations needed to partition into its constituent elements the difference between the U.S. and U.K. ratios of energy consumed for passenger transport relative to GDP. As noted in the tabulation above, these two figures are, respectively, 217 and 82 tons of oil equivalent per million dollars of GDP. The same figures appear in the last line and last column of, respectively, panels A and F in table E-1. (Note, however, that in panel F, all the percentages and ratios are the actual U.K. figures as also shown in table 5-3; but the absolute energy consumption and passenger-mile figures are hypothetical, having been scaled up to the U.S. level of GDP.)

In this exercise, the factors into which the passenger transport energy consumption/GDP ratio are decomposed are: modal mix, that is, the proportion of nationwide passenger travel accounted for by the different means of transport; the "activity mix," by which we mean the volume of travel (in passenger-miles) relative to GDP; and energy intensity, the amount of energy consumed per passenger-mile for a given mode. Even though the terms are often used interchangeably, "energy intensity" seems preferable to "energy efficiency" as a characterization of the final ratio, insofar as it reflects a strictly neutral rather than normative connotation of performance. (The ratio, after all, reflects more than a purely thermodynamic, energy conversion process.)

Turning, once again, to table E-1, we find in panel B a calculation of the U.S. passenger transport energy/GDP ratios on the basis of the British modal mix, other factors unchanged. In panel C, the British activity mix (that is, the passenger-mile/GDP ratio) is adopted, other factors re-

TABLE E-1 *Actual and Hypothetical Passenger Transport Energy Consumption, United States–United Kingdom, 1972*

Passenger travel modes	Energy consumption		Passenger-miles		Energy consumption (tons oil equiv. per 100 thousand passenger-miles)	Passenger-miles per $ thousand GDP	Energy/GDP ratio (tons oil equiv. per $ million GDP)
	Million tons oil equiv.	Percentage	Billions	Percentage			
Panel A. U.S. actual							
Cars	226.92	88.5	2,171.3	92.1	10.450	1,842	192.6
Buses	1.74	0.7	49.0	2.1	3.550	42	1.5
Rail	0.84	0.3	17.8	0.8	4.758	15	0.7
Air	26.78	10.4	119.3	5.1	22.450	101	22.7
Total	256.28	100.0	2,357.3	100.0	10.872	2,000	217.5
Panel B. Assuming total actual U.S. passenger-miles distributed at U.K. modal mix, and U.S. intensity							
Cars	196.33	90.1	1,878.8	79.7	10.450	1,594	166.6
Buses	10.29	4.7	289.9	12.3	3.550	246	8.7
Rail	8.53	3.9	179.2	7.6	4.758	152	7.2
Air	2.65	1.2	11.8	0.5	22.450	10	2.2
Total	217.80	100.0	2,357.3	100.0	9.239	2,000	184.8
Panel C. Assuming U.S. passenger-mile modal mix at U.K. total passenger-mile/GDP ratio, and U.S. intensity							
Cars	166.34	88.5	1,592.2	92.1	10.450	1,351	141.1
Buses	1.32	0.7	36.3	2.1	3.550	31	1.1
Rail	0.56	0.3	13.8	0.8	4.758	12	0.5
Air	19.55	10.4	88.2	5.1	22.450	75	16.6
Total	187.96	100.0	1,728.8	100.0	10.872	1.467	159.5

Panel D. Assuming both U.K. modal mix and U.K. total passenger-mile/GDP ratio, and U.S. intensity

Cars	143.97	90.2	1,377.7	79.7	10.450	1,169	122.2
Buses	7.53	4.7	212.1	12.3	3.550	180	6.4
Rail	6.22	3.9	130.8	7.6	4.758	111	5.2
Air	1.84	1.2	8.2	0.5	22.450	7	1.6
Total	159.56	100.0	1,728.8	100.0	9.259	1,467	135.4

Panel E. Assuming U.K. intensity, and both U.S. modal mix and total passenger-mile/GDP ratio

Cars	124.52	76.3	2,171.3	92.1	5.735	1,842	105.7
Buses	1.78	1.1	49.0	2.1	3.632	42	1.5
Rail	1.05	0.6	17.8	0.8	5.922	15	0.9
Air	35.79	21.9	119.3	5.1	30.000	101	30.4
Total	163.14	100.0	2,357.3	100.0	6.921	2,000	138.4

Panel F. Assuming U.K. intensity, U.K. modal mix, and U.K. total passenger-mile/GDP ratio

Cars	79.01	81.6	1,377.7	79.7	5.735	1,169	67.1
Buses	0.77	7.9	212.1	12.3	3.632	180	6.5
Rail	0.77	8.0	130.8	7.6	5.922	111	6.6
Air	0.25	2.5	8.2	0.5	30.000	7	2.1
Total	96.90	100.0	1,728.8	100.0	5.605	1,467	82.3

maining unchanged. In panel D, both the modal mix and the activity mix are varied, the U.S. intensity (that is, the energy consumption per passenger-mile ratio) being retained. Panel E shows what happens when the U.K. intensity is coupled with the U.S. modal and activity mixes.

The final column of each panel shows the energy/GDP ratios (for given modes and for all passenger transport combined) that result from these hypothetical calculations. It is these figures that form the basis of table E-2. This table that, very loosely speaking, tells us how much less energy the U.S. passenger transport systems would require if we did things the "British way," provides a summary sorting-out of the factors responsible for the difference in the U.S.–U.K. passenger transport energy/GDP ratios. For example, the figure in the first column, first line, of table E-2 indicates that the United States consumes 26.0 more tons of oil equivalent per million dollars of GDP for automotive use than it would if it had the British modal mix pattern. The figure of 26.0 was obtained by going to the last column of table E-1 and subtracting the first line of panel B from the first line of panel A.

The last line in table E-2 (which is what is summarized for all binary comparisons in table 5-4 of chapter 5) tells us that the entire U.S.–U.K. difference is split about evenly between the intensity phenomenon, on the one hand, and the two mix factors, on the other. Within the mix factors, the activity mix seems to be more important than the modal mix. That is, the fact that Americans travel more relative to their income compared to the British accounts for a larger portion of our greater passenger transport energy use than does the fact that our modal mix differs from Britain's.

Note the columns marked "interaction" in table E-2. These columns arise because we have partitioned the U.S.–U.K. passenger transport energy/GDP difference into multiplicative, rather than additive, factors. This is a familiar problem that is often encountered in the simplification of quantitative relationships. The easiest way to illustrate the matter is by means of the schematic representation below.

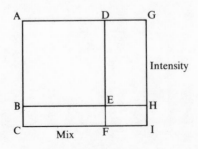

TABLE E-2 *Analytical Breakdown of U.S.–U.K. Differences in Passenger Transport Energy/GDP Ratios, 1972*
(tons oil equivalent per $ million GDP)

| Modes | Due to mix factors | | | | Due to intensity factors | Due to interaction between total mix and intensity | Total actual U.S.–U.K. difference |
	Modal mix effect	Activity mix effect	Interaction	Total mix effect			
Cars	26.0	51.5	−7.1	70.4	86.9	−31.8	125.5
Buses	−7.2	0.4	1.9	−4.9	0.0	0.1	−5.0
Trains	−6.5	0.2	1.8	−4.5	−0.2	−1.3	−5.9
Air	20.5	6.1	−5.5	21.1	−7.7	7.2	20.6
Total	32.7	58.0	−8.6	82.1	79.1	−26.0	135.2

Assume, for present purposes, that mix times efficiency equals energy consumption and that the larger square (ACIG) represents the United States and the smaller (inset) square (ABED), the United Kingdom. If we hold U.S. intensity constant but adopt U.K. mix, we obtain the area ACFD; that is, we subtract the area DFIG. If we hold U.S. mix constant and adopt U.K. intensity, we obtain the area ABHG; that is, we subtract the area BCIH. But note that we have now subtracted the area EFIH twice. Thus, if we want to proceed from the larger to smaller square in the manner just described, we will have to cancel one of the EFIHs—hence, the negatives in the last line of table E-2. (If we had proceeded in a reverse fashion—by going from the smaller to larger square—we would have missed EFIH altogether and would have had to add it on.)

For convenience of representation—although at the price of some distortion—a simplification sometimes employed is to remove the interaction effect by distributing it proportionally among the remaining constituent factors. If we had done that, the last row of table E-2 would appear as follows:

	Interaction	
	In	Out
	(tons oil equiv. per $ million GDP)	
Modal mix effect	32.7	26.1
Activity mix effect	58.0	46.2
Interaction	−8.6	—
Total mix effect	82.1	72.3
Due to intensity factors	79.1	62.9
Due to interaction between total mix and efficiency	−20.6	—
Total actual U.S.–U.K. difference	135.2	135.2

After removing the interaction effect, and thereby obtaining an additive total, one could then break down the total difference into the following percentages.

Due to modal mix	19.3
Due to activity mix	34.2
Due to intensity	46.5
	100.0

Estimation of Energy Price Indexes

To assess the influence of prices on energy consumption it was first necessary to develop energy price indexes for each of our countries. Separate indexes were calculated for each of the principal sectors, as some countries have low energy prices in one sector but high prices in another. Note that, in principle, all prices are inclusive of taxes.

The estimation of sectoral price indexes involved several stages. First, collecting raw data for each country; second, putting this data on a comparable basis with respect to quantity and quality of the energy involved; third, converting comparable quantity data expressed in national currencies to a common unit of currency; and finally, in those sectors where more than one fuel is used—household-commercial and industrial—the construction of sectoral price indexes based on the weight of the quantities of the different fuels used within the sector. There were difficulties of interpretation and estimation at all stages of the calculation. Consequently, the price indexes must be interpreted very broadly as approximate levels of energy prices relative to those of the United States.

The Basic Data and Their Comparability

Basic data on prices were gathered from a variety of national and international sources. For price data applying to the individual country, see the sources footnoted in each chapter. International sources that include data for several, though not necessarily all, of our countries are:

- European Communities, *Energy Statistics Yearbook, 1969–72* (Luxembourg, European Communities Statistical Office [EUROSTAT], 1974).

- European Communities, *Energy Statistics: Prices of Fuel Oils, 1960–74,* Special Number 1, 1974 (Luxembourg, European Communities Statistical Office [EUROSTAT]).

- European Communities, *Energy Statistics: A Comparison of Fuel Prices: Oil, Coal, Gas, 1955–70,* Special Number 2, 1974 (Luxembourg, European Communities Statistical Office [EUROSTAT]).

- European Communities, *Energy Statistics: Prix du Charbon (Coal Prices), 1955–70,* Supplement Number 1-2, 1973 (Luxembourg, European Communities Statistical Office [EUROSTAT]).

- U.S. Department of the Interior, Bureau of Mines, *International Petroleum Annual, 1972* (1974).

- U.S. Department of Commerce, Domestic and International Business Administration, Office of Economic Research, "Selling Prices and Taxes on Motor Gasoline, Diesel Fuel and Heating Oil in Selected Countries as of October 1974," mimeo.

Help with price data was also received from OECD and the World Bank. From these sources data on the following fuels were compiled:

- for the transport sector—regular grade gasoline
- for the industrial sector—heavy fuel oil, natural gas, electricity, coal.
- for the household-commercial sector—electricity, gas, coal products, heating oil, and (in some cases) liquid gases and kerosine.

Although this list is not exhaustive it does cover most of the forms of fuel and energy used in our countries. Particular efforts were made to secure good comparable price data for those fuels—electricity, gas, heating oil, and fuel oil—that are used extensively in all countries. For those used in quantity only in a few countries (kerosine in Japan and coal in the U.K. household-commercial sector), efforts were limited to securing price data for the country concerned (to the detriment, as will be seen later, of calculating pure price indexes).

In principle, the data apply to average 1972 prices. In fact, some of our sources refer to the beginning of 1972, and some to mid-1972. Fortunately for our purposes, there does not seem to have been much movement in energy prices in our countries over this period so that there is some justification for considering these prices as representative of a yearly average. For those fuels—electricity and natural gas—where a range of prices exists, depending on the quantity purchased, there was clearly some problem of selecting an average or typical single rate. This problem was particularly acute for electricity in the household-commercial sector, where prices fall off very sharply as consumption increases. Wherever possible the selection of an average price was guided by average revenue data. And where such data were not available, the rate selected was chosen on the basis of the relationship between average revenue and consumption ranges noted in other similar countries.

Conversion to a Common Unit of Currency

The next step was to convert the price per kilocalorie, at this point expressed in national currencies, into a common unit of currency, the U.S. dollar. It will be recalled from appendix A that purchasing-power-parity rates of exchange were preferred to market rates for the conversion of GDP expressed in national currencies into U.S. dollars. But the parities referred to in that appendix apply to national output on the aggregate level (GDP) and could therefore be misleading if used in the deflation of values associated with or contained within sectoral components rather than aggregates. For this reason, special "consumption" purchasing power parities were calculated for use in deflating those energy prices closely associated with the consumption sector (household energy prices and gasoline prices). The other prices, those applying to industrial uses, were deflated by the GDP purchasing power parities, which in the absence of any satisfactory "industrial" purchasing power parity were felt to be more appropriate deflaters than the consumption parity.

The estimation of the consumption purchasing power parities followed the same procedure as that outlined in appendix A. First, 1970 purchasing power parities (Ideal) applying to the consumption sector only were taken from tables 13.2, 13.3, 13.6, 13.7, and 13.9 of the *ICP Report* for the following countries: France, West Germany, Italy, United Kingdom, and Japan. Consumption parities were interpolated for the countries not covered in the *ICP Report:* Canada, Sweden, and the Netherlands—following methods described in appendix A. These data were then adjusted upward by the rise in each country's consumer price index of 1970–72 relative to the movement in U.S. consumer prices. The results of this operation are given in table F-1. In general there is little difference between the GDP and consumption purchasing power parities. Where there is a difference, it is usually in the direction of more units of foreign currency being equivalent to a dollar in the consumption sector than for GDP as a whole.

The resulting 1972 consumption purchasing power parities were used to deflate gasoline and household-commercial energy prices expressed in national currencies. The GDP parities were used to deflate industrial prices. The resulting price data, expressed in U.S. dollars per kilocalorie,

TABLE F-1 *Consumption and GDP Purchasing Power Parities (Ideal Weights), 1972*

Purchasing power parity	Canada	France	W. Germany	Italy	Netherlands	U.K.	Sweden	Japan
	C$/$	Fr/$	DM/$	L/$	g/$	£/$	Kr/$	Y/$
Consumption purchasing power parities	1.003	4.82	3.44	4.93	3.06	0.329	5.04	247
GDP purchasing power parities	1.007	4.65	3.37	4.83	3.00	0.322	4.90	250

Note: Line 1 derived from updated *ICP Report* parities. Line 2, from appendix A.

are in table F-2. Despite the quality of some of the price data, and differences in interpretation inherent in the choice of price deflators, some broad conclusions can be drawn.

First, except in the case of petroleum products, the United States has in general significantly lower energy prices than any other country. Second, there are, nonetheless, significant differences between sectors. Within transport (using gasoline prices as representative of transport energy prices) United States and Canada have substantially lower prices than all the other countries. But within the household-commercial and industrial sectors, the differences are narrowed.

Calculation of Sectoral Indexes

The possibility of substituting one fuel for another within a given sector brings us to the final section of this appendix—the calculation of price indexes for those sectors (industrial and household-commercial) that use more than one fuel. In these sectors, fuel mixes between countries differ considerably, and to the extent that prices of the fuels vary, the weighted average for the sector will depend on the extent to which consumption is concentrated in the more or the less expensive fuels.

Two sectoral indexes were calculated—a pure price index and a unit value index. Both are weighted by the relative importance of the different fuels in total sectoral consumption. The difference between them depends on the type of weighting. The pure price index is based on the geometric average of the prices and quantity shares of the United States and each other country in turn. The unit value index weights a country's prices only by the quantities consumed in that country. This gives the average or unit cost per kilocalorie in any one country.

The pure price index is theoretically superior but it has the disadvantage in this case of requiring more data than were available. Thus, for each country the pure price index calls for price and quantity data for all fuels used by the United States. This is particularly troublesome in the case of fuels which are widely used in the United States but not in other countries. It is frequently very difficult to find price data for a fuel which is little used in a country. Furthermore, even when such data are available, they can be misleading. Japan, for example, uses a very small amount of very high-priced gas in industry, presumably only when no other fuel will serve. The multiplication of this exceptionally high price by the sizable proportion of industrial consumption of gas in the United

TABLE F-2 *Energy Prices Deflated by Purchasing Power Parities, 1972*
(U.S. cents or dollars)

Sectors	U.S.	Canada	France	W. Germany	Italy	Netherlands	U.K.	Sweden	Japan
Transport sector [a]									
Regular gasoline (cents per liter)	9.8	7.2	23.1	17.1	30.8	22.7	22.5	19.7	19.7
Household-commercial sector [a] (dollars per million Kcal)									
Electricity	26.6	32.3	59.7	49.9	51.9	35.3	35.0	25.2	66.6
Gas	4.7	3.8	13.8	12.9	8.0	7.1	10.0	b	b
Coal	b	b	11.2	10.4	b	b	10.7	b	b
Heating oil	3.1	4.0	5.1	3.7	4.3	4.3	5.9	5.2	4.8
Kerosine	b	b	b	b	b	8.5	b	b	9.7
Industrial sector [c] (dollars per million Kcal)									
Electricity	11.0	17.7	21.9	28.7	27.2	26.1	24.9	15.0	25.4
Gas	1.7	1.7	3.1	3.3	3.0	2.1	2.3	b	b
Coal	1.8	1.9	3.9	3.9	4.7	3.4	3.9	3.6	3.6
Petroleum products	2.8	2.2	3.7	2.9	3.5	2.5	4.3	2.7	3.2

[a] Prices deflated by consumption purchasing-power-parity rates of exchange.
[b] Negligible consumption.
[c] Prices deflated by GDP rates of exchange.

States (50 percent) would drive up the pure price index of Japan to unrealistic levels.

The unit value index, however, requires price data only for those fuels which are substantially consumed within a country—data which are, in any event, invariably easier to come by. Though less satisfactory as an indication of prices between countries, this index provides a better indication of the varying costs of a given amount of fuel and power between countries.

These indexes are given in table F-3, along with an index of gasoline prices representing the price of energy used in transport. For the industrial sector, the two indexes differ little. The reason for this similarity is, first, the similarity of prices of fuels other than electricity within each country; and second, some similarity in electricity shares between countries. The biggest divergence between the two indexes occurs in the case of Sweden, which uses a much higher proportion of electricity in industry than any other country, including the United States.

In the household-commercial sector, the difference between the two indexes is more pronounced. The pure price index for almost all countries is much higher than the unit value index, largely because of the larger U.S. share of electricity in the household-commercial sector. The greater consumption by the European countries and Japan of non-electricity fuels, which are much cheaper on a thermal basis than electricity, reduces the unit cost of their fuel by more than might be indicated by the differences in prices of the individual fuels.

The individual sectoral prices are analyzed in more detail in the sectoral analyses in chapters 4, 5, and 6. In brief, the indexes show:

For transport, the index of 1972 gasoline prices indicates three groups of countries: United States and Canada as having low prices; Italy high prices, at 314; and the other countries falling in between.

For household fuel and power, the United States and Canada appear to have the lowest prices (as measured by the pure price index), France and Japan, the highest (about double the level of the low-price countries). However, there is clearly some tendency in many countries to consume the lowest price fuel. Thus, as noted previously, the range of the unit value index is much narrower relative to the United States than that of the pure price index. Sweden, Netherlands, and the United States now become lower unit value countries and the United Kingdom and Japan, although still remaining the highest, are about 70 percent higher than the United States, compared with 100 percent with the pure price index.

TABLE F-3 *Energy Price Indexes, 1972*
(United States = 100)

Indexes	Canada	France	W. Germany	Italy	Netherlands	U.K.	Sweden	Japan
Transport sector								
Gasoline	114.4	235.2	174.9	314.4	231.9	229.5	201.3	200.6
Household-commercial sector								
Pure price index	114.6	205.3	180.9	180.2	150.1	165.4	147.0	190.4
Unit value index	109.5	141.1	131.2	132.0	104.2	175.4	110.0	170.0
Industrial sector								
Pure price index	131.5	182.6	210.3	204.8	188.3	193.0	138.7	203.1
Unit value index	158.4	185.6	202.8	206.1	167.8	183.5	187.2	225.1

For industrial use, energy prices in most countries are generally about double the U.S. levels. The exceptions are Canada, some 50 percent higher, and, somewhat ambiguously, both Sweden and Netherlands, which can be some 50 percent higher than the United States, depending on the weighting system used.

In the interpretation of these indexes, several qualifications must be borne in mind. First, the price indexes, particularly for the household-commercial sector, are highly sensitive to the prices of heating oil and electricity, and data on both are uncertain. Electricity prices could vary widely (perhaps as much as 30 percent) from the rates used here. Second, the effect of converting energy prices by purchasing power parities rather than market exchange rates is to yield prices that are higher in relation to the U.S. levels. And, third, some results are highly dependent on the weighting system used in calculating sectoral indexes.

For all these reasons the price indexes will support only the broadest conclusions. Nonetheless, it can be stated with some confidence that, even taking into account the most extreme examples of the sensitivities referred to above, the United States would still emerge as a country with low energy costs. Moreover, the difference in prices between the United States and other countries is greatest in transport, somewhat less in industry, and least of all in the household-commercial sector.

Input–Output Methodology for Estimating Effect of Final-demand Mix on Aggregate Energy/GDP Ratios

THE ENERGY/GROSS DOMESTIC PRODUCT (energy/GDP) ratios of other countries can vary from that of United States because the mix of final goods and services is different, the direct and indirect energy content of individual final products is different, or both. A useful analytical tool for determining how much of the difference in energy/GDP ratios of other countries relative to United States is due to each of these factors involves the use of input–output tables. This approach is similar to the conventional procedure used in input–output analysis to partition the change over time in the industrial composition of a country into (1) the final-demand mix effect and (2) the effect of changes in input–output coefficients. The procedure involves the use of input–output tables for each of the two periods being compared. In this procedure, the effect of final-demand mix is calculated, holding constant the input–output coefficients and then computing the effect of changes in input–output coefficients, holding constant the final-demand mix. The sum of the final demand and coefficient change effect does not exhaust the total change, because the interaction between the two factors accounts for part of the total change. The interaction effect can be measured and then either shown separately as a third factor or distributed between the two variables so as to exhaust the total change.

A similar procedure can be used to partition the difference in energy/GDP ratios between United States and other countries. When comparing countries, usually only part of the partitioning analysis can be done, because input–output tables for the individual countries are generally not compatible in terms of classification systems, input–output accounting conventions, and sets of prices. Therefore, the input–output analysis developed in this report was limited to determining how much of the difference in aggregate energy/GDP ratios of other countries relative to that of United States is attributable to the differences in final-demand mix, assuming U.S. input–output coefficients for all countries being compared.[1]

This appendix describes the methods used to develop the "bills of goods" for final-demand components of the countries involved in the comparison, the treatment of the various bills of goods so as to be compatible with the 1970 U.S. input–output table, and the conversion of the dollar value of energy consumption generated in the course of these computations into British thermal units (Btus).[2]

Derivation of Final-demand Bills of Goods

The basic data source for estimating the final-demand bills of goods was the United Nations-sponsored report on Phase One of the International Comparisons of Gross Product and Purchasing Power.[3] The *ICP Report,* as it has been referred to throughout this study, provides data for five of the countries (in addition to United States) that are included in the present report: France, West Germany, Italy, United Kingdom, and Japan. Data are provided for all major components of gross domestic product, but detailed estimates are not given for all the major components. The detailed estimates are limited to consumption purchases, construction, and producer durables. For this reason the input–output analysis was limited to these components of GDP.

[1] It should be noted that the decomposition of differences in energy/GDP ratios into the mix versus energy intensity factors has different meanings depending on the particular approach used. In the input–output approach, "mix" refers to the differences in mix of expenditures for final goods and services (including all direct purchases of energy by final demand), and energy intensity reflects differences in all input–output coefficients (including nonenergy as well as energy coefficients). In the sector-by-sector analysis based on energy balance tables, "mix" refers to the relative importance of sector output (including sector component detail), and implicitly includes both the mix of final-demand expenditures (excluding the energy intensity portion of direct purchases of energy by final demand) and the nonenergy input–ouput coefficients. Energy intensity refers, in concept, to direct energy input–output coefficients and the energy intensity portion of direct energy purchases by final demand.

[2] The development of (1) the final-demand bills of goods, (2) the multiplication of the bills of goods against the U.S. table of input–output coefficients (covering direct and indirect coefficients as reflected in the input–output total requirements table) in order to derive estimates, in value terms, of energy consumption and (3) the conversion of energy consumption into Btu terms were done by the Office of Economic Growth, Bureau of Labor Statistics, U.S. Department of Labor. The Btu/dollar conversion factors were supplied by RFF.

[3] Irving B. Kravis, Zoltan Kenessey, Alan Heston, and Robert Summers, *A System of International Comparisons of Gross Product and Purchasing Power* (Baltimore, published for the World Bank by The Johns Hopkins University Press, 1975).

For these components, the *ICP Report* gives detailed data on per capita purchases, in own-country currencies, and purchasing power parities (which are used to convert other-country purchases into 1970 U.S. prices).[4] The detailed data on per capita expenditures, in U.S. prices for all countries were multiplied by population estimates for each country to derive estimates of expenditures for each item of final demand covered in the input–output analysis.

The sums of the detailed items of consumption and fixed investment purchases (in U.S. prices) used in the input–output analysis did not exactly match the total purchases for the components derived by direct use of the aggregate purchasing power parities for these categories. This was due to the fact that, in the case of consumption purchases, the item "expenditures for residents abroad" was excluded and also the purchasing power parities used to fill in the gap for some missing consumption items were different than those implicit in the aggregate consumption purchasing power parity. This latter factor also explains the disparity between the estimates used in the input–output analysis for construction and producer durables and the estimates derived from the direct use of the aggregate purchasing power parities for these categories. The differences between the two estimates of total purchases for the various components were small and did not significantly affect the estimates of the effect of demand mix on the aggregate Btu/dollar ratios.

Having developed estimates of purchases in U.S. prices of the detailed items of consumption, construction, and producer-durable equipment, the next major step was to convert these into categories that matched the U.S. input–output table for 1970 that was developed by the Bureau of Labor Statistics. The 1970 input–output table is an updated version of the 1963 table developed by the Bureau of Economic Analysis, U.S. Department of Commerce, as part of the official national income and product accounts. The 1970 table involves a consolidation of the original detailed input–output classification system into 135 industries. The updated 1970 table is in 1963 prices, consistent with the table from which it is derived.

The conversion of the item detail in the ICP accounts into final-demand bills of goods that are consistent with the 1970 table was done in several

---

[4] The data on per capita expenditures and purchasing power parities in the *ICP Report* are in the appendix to chapter 13, and detailed binary tables are on pp. 195–227. Population estimates are given in table 1.1 on p. 6. For some of the items, data on purchasing power parities are not shown in the detailed binary tables. In those instances, estimates were developed from information provided in appendix table 14.3, "Purchasing Power Parities per U.S. Dollar, Nine Countries, and International Prices, 1970" on pp. 248–250 or from per capita quantity indexes shown in the detailed binary tables in the appendix of the *ICP Report*.

stages. First, the ICP detailed items were grouped into categories that matched those in bridge tables used to convert final-demand component detail to input–output industry detail. Second, the ICP category estimates were deflated from 1970 to 1963 prices to be consistent with the prices of the input–output table. Third, the estimates for United States were adjusted to match the actual levels for these groups in the 1970 table. This was done only for the United States, because this was the only country for which ICP estimates could be related to comparable input–output estimates. Fourth, the final-demand estimates in 1963 prices were multiplied by the appropriate bridge tables to derive separate input–output bills of goods for each major component of final demand, that is, consumption, construction, and producer-durable equipment.

The degree of aggregation to match the categories of the final-demand bridge tables varied among the final-demand components. For consumption expenditures, 109 of the 110 consumption items were grouped into 61 product groups. The remaining consumption item (expenditures of residents abroad) was excluded, because the conversion of such expenditures into domestic energy consumption attributable to such expenditures was not meaningful.

It should be noted that in the ICP accounts, the consumption component includes both household consumption expenditures and purchases by government for medical, educational, and recreational functions. In concept, the purchases by households and government for similar functions should be related to separate bridge tables, because the allocation of such purchases to individual input–output industries, particularly trade industries, may be different for government purchases than they are for personal consumption expenditures for these categories. In practice, it is difficult to develop separate bridge tables that adequately reflect such differences. In the case of consumption expenditures, an effort was made to develop separate bridge tables for the specific categories of government purchases transferred in the ICP accounts from government to the consumption component of GDP. Information on such transfers is given in the *ICP Report*.[5] Of the 61 product groups in the bridge table for consumption expenditures, five special bridge tables were developed for specific government expenditures. The government functions treated separately were education (compensation, and other purchases), drugs, medical supplies, and therapeutic equipment.

The detail on fixed investment included fourteen types of construction activity and twenty-two producer-durable equipment items. The fourteen

---

[5] *ICP Report,* table 13.15 "Government Components of Final Consumption Expenditures" on p. 186.

types of construction were aggregated into the four types shown separately in the input–output table. Classifying the construction activities of other countries into the four construction activities in the U.S. input–output table assumes the same product and material input mix for each of the four types of construction as exists in United States. For producer-durable equipment, the bridge table that was developed matched the twenty-two items so that no aggregation of these items was required. The ICP fixed investment components include both private and government non-defense expenditures for construction and producer durables. In concept, this would suggest that separate bridge tables should have been developed for the government purchases of these items, but unfortunately the *ICP Report* did not give data similar to that for government purchases transferred to consumption, which would show the distribution between private and public expenditures for fixed investment. As a result, the bridge tables, which are based on U.S. private purchases, were applied to the combined purchases.

Because the bridge tables played an important role in converting the ICP final-demand categories into input–output bills of goods, their function will be discussed in a little more detail. The main purpose of the bridge table is to distribute the initial estimates of final-demand purchases into the component input–output industries that produce, transport, and distribute the final products and services which make up the bundle of items included in the ICP final-demand groups. There are two aspects to this distribution into component industries. First, an ICP final demand-category may be "produced" by more than one industry. For example, household purchase of fish includes fresh fish, a product of the "forest and fishery" industry, canned and frozen fish, products of the "food products" industry, and imported fish of all types which in the U.S. input–output system are considered to be direct purchases from other countries ("directly allocated imports"). In addition, the bridge table also distributes the total value of purchases for each final-demand category between the amount received by each of the producing industries and the amount received by various "margin" industries, that is, the industries that provide transportation, trade, and insurance services. The transportation costs are further distributed by mode of transportation, that is, railroads, trucking, water, air, and pipeline.

Multiplying the ICP final-demand purchases for each component of GDP included in the input–output analysis by the related bridge tables distributes these purchases, classified by category detail in the ICP tables, into the bills of goods classified by the 135 input–output industries.

The purchases from each of the industries are stated in producers' value, but since the costs of transportation, distribution, and insurance are also treated as separate industries in the bills of goods, the sum of all purchases from the producing and margin industries is equal to the total purchases by each component of final demand.

The separate bill of goods for each final-demand component (consumption, construction, and producer-durable equipment) was then multiplied by the input–output coefficients covering direct and indirect requirements to provide estimates of the output from each energy-producing industry required directly or indirectly to meet the stipulated bill of goods.

It should be noted that since the bridge tables used in developing the bills of goods are based on the U.S. composition of production, transportation, and distribution industries for specific items of final demand, then part of the difference between the energy/GDP ratios of other countries relative to that of United States may lie in the fact that the bridge tables of other countries may be sufficiently different from that of the United States as to affect the aggregate Btu/GDP ratios. In this sense, the U.S. distribution factors built into the bridge tables are analogous to the U.S. input–output coefficients, which are assumed for the purpose of this analysis to hold for all the countries included in the analysis.

Conversion of Energy Content of Final-demand Components into British Thermal Units

Having developed estimates of the direct and indirect energy required to produce, transport, and distribute the goods and services in each component of final demand, the next major step was to convert these estimates of energy requirements stated in U.S. dollars, into thermal units (Btus). There were two major problems involved in converting dollar estimates of energy into Btus that are consistent with Bureau of Mines' data on U.S. domestic energy consumption. The first problem was how to avoid double-counting primary and secondary sources of energy embodied in the final product. The second was to develop conversion factors that take account of the fact that different categories of purchasers pay different prices for energy and that the same conversion factor cannot be used for purchases of energy by households as for industrial and commercial purchases.

The first problem was how to convert the output of the energy industries from dollars to Btus without double-counting energy (for example—

how to avoid counting the electricity generated from coal as well as coal in its primary form). In the Bureau of Mines' energy balance accounts, the basic energy sources are coal, refined petroleum products plus natural gas liquids, dry natural gas, and electric power (limited to hydropower and nuclear power). Note that crude petroleum is excluded entirely in order not to duplicate the Btu content of refined petroleum products. Electric power generated from coal, petroleum, or natural gas is also excluded so as not to duplicate the Btus already covered by these three types of energy as primary sources.

The method used in this report to convert dollar outputs of energy industries without double-counting was to follow the categories of basic energy sources used by the Bureau of Mines and relate them to the corresponding input–output energy industries. The problem itself was largely confined to two industries—crude petroleum and electric utilities. As noted previously, the Bureau of Mines excludes the Btu content of crude petroleum, preferring to use the refined petroleum stage as the basis for estimating the Btu content of petroleum. Since almost all crude petroleum has to be refined to be used, this presents no problem if we follow this procedure and also convert dollars to Btus at the refined petroleum stage. It also has the advantage of bypassing the problem of distributing the output of the crude petroleum and natural gas extraction industry between the two major components. The petroleum is counted at the refinery stage, the natural gas is counted at the gas utility stage. The natural gas liquids (products of the crude petroleum and natural gas industry) are added to the output of the petroleum refining industry so that eliminating the crude petroleum and natural gas mining industry from the calculations does not mean that any energy will be left out, but that it will be picked up at a later stage.

The conversion of the output of the electric utility industry into Btus presents a special problem because counting the Btu content of the total output of the electric utility industry would involve double-counting the Btu content of coal, refined petroleum, and natural gas, which are used to generate electric power. On the other hand, electric power generated as hydropower and nuclear power does not involve double-counting and must be converted into Btus if the energy content of final-demand bills of goods is to be complete, covering all forms of unduplicated energy. The solution used was to relate the Btu content of hydropower and nuclear power (both computed on the basis of the heat rate of electric power generated from fossil fuels) to the value of the total output of the electric power industry on the assumption that the use of nonfossil-fueled electric power by consuming industries and final-demand categories is in proportion to their use of total electric power. It has the effect of scaling down purchases of total electric power generated by each bill of goods to the proportionate share contributed by hydroelectric and nuclear power, based on U.S. proportions.

The other major problem encountered in converting the energy content of final-demand components into Btus was that the Btu/dollar conversion factor developed for each energy industry is an average for the entire industry, covering substantial variation within the average because of differential prices paid by different categories of users and differences in energy content of specific products within industries. In order to take account of the major distortion which might result from using the average industry conversion factor for all categories of users, a separate set of conversion factors was used for direct purchases of energy by households.[6] It was this category of final-demand purchases, covering individual fuel and power items for household operations and gasoline, and oil and grease for personal auto transportation, that was considered to be the major potential source of distortion.

The special set of conversion factors for direct energy purchases by households was used to derive the Btu content of such purchases. The Btus and value of household energy purchases were then deducted from the aggregate Btus, derived on the basis of the initial set of Btu/dollar conversion factors, and aggregate final demand. This yielded a "residual" estimate of Btu/final demand estimates covering all final demand, excluding direct energy purchases by households. The difference between the residual estimate and the estimate based on the earlier average Btu/dollar conversion factors was used as an adjustment factor for modifying the first set of Btu/final demand ratios for the various components of final demand, excluding direct energy purchases by households.

Finally, it should be noted that although the detailed input–output bills of goods were developed in 1963 prices in order to be consistent with the price level of the input–output table, at a final stage the estimates of the Btu content of final-demand components were related to the final-demand purchases in 1970 dollars in order to reflect the price level used in the initial conversion of other countries' final-demand purchases to a common price level—1970 U.S. prices.

---

[6] The conversion factors for direct energy purchases for personal consumption expenditures were based on: Bruce Hannon and Nadine Abbot, *Energy and Employment Impacts of Final Demand Activities, 1963* (Urbana, Ill. Center for Advanced Computation, University of Illinois, 1974) document 128.

Index

Index

279